MATHEMATICS AND ITS TEACHING IN THE MUSLIM WORLD

SERIES ON MATHEMATICS EDUCATION

Mathematics education is a field of active research in the past few decades. Plenty of important and valuable research results were published. The series of monographs is to capture those output in book form. The series is to serve as a record for the research done and to be used as references for further research. The themes/topics may include the new maths forms, modeling and applications, proof and proving, amongst several others.

Published

Vol. 16 *Mathematical Outreach: Explorations in Social Justice Around the Globe*
edited by Hector Rosario

Vol. 15 *Mathematics and its Teaching in the Asia-Pacific Region*
edited by John Mack and Bruce R Vogeli

Vol. 14 *Mathematics and its Teaching in the Muslim World*
edited by Bruce R Vogeli and Mohamed El Tom

Vol. 13 *K-12 Mathematics Education in Israel: Issues and Innovations*
edited by Nitsa Movshovitz-Hadar

Vol. 12 *Special Secondary Schools for the Mathematically Talented: An International Panorama*
edited by Bruce R Vogeli

Vol. 11 *Mathematics Education in Korea*
Volume 2: Contemporary Trends in Researches in Korea
edited by Jinho Kim, Inki Han, Mangoo Park and Joongkwoen Lee

Vol. 10 *Mathematics and Its Teaching in the Southern Americas: with An Introduction by Ubiratan D'Ambrosio*
edited by Héctor Rosario, Patrick Scott and Bruce R Vogeli

Vol. 9 *Primary Mathematics Standards for Pre-Service Teachers in Chile: A Resource Book for Teachers and Educators*
by Patricio Felmer, Renato Lewin, Salomé Martínez, Cristián Reyes, Leonor Varas, Eugenio Chandía, Pablo Dartnell, Alejandro López, Claudio Martínez, Arturo Mena, Andrés Ortíz, Gloria Schwarze and Pierina Zanocco

Vol. 8 *Mathematical Modelling: From Theory to Practice*
edited by Ngan Hoe Lee and Dawn Kit Ee Ng

For the complete list of titles in this series, please go to
http://www.worldscientific.com/series/sme

Series on Mathematics Education Vol. 14

MATHEMATICS AND ITS TEACHING IN THE MUSLIM WORLD

Edited by

Bruce R Vogeli
Columbia University, USA

Mohamed E A El Tom
Ministry of Education, Sudan

NEW JERSEY • LONDON • SINGAPORE • BEIJING • SHANGHAI • HONG KONG • TAIPEI • CHENNAI • TOKYO

Published by

World Scientific Publishing Co. Pte. Ltd.
5 Toh Tuck Link, Singapore 596224
USA office: 27 Warren Street, Suite 401-402, Hackensack, NJ 07601
UK office: 57 Shelton Street, Covent Garden, London WC2H 9HE

Library of Congress Cataloging-in-Publication Data
Names: Vogeli, Bruce R. (Bruce Ramon), editor. | Tom, M. E. A., 1941– editor.
Title: Mathematics and its teaching in the Muslim world / edited by Bruce R. Vogeli, Columbia University, Mohamed El Tom, Khartoum Technical University, Sudan.
Description: First. | New Jersey : World Scientific Publishing Co., [2020] | Series: Series on mathematics education, 1793-1150 ; vol. 14 | Includes bibliographical references.
Identifiers: LCCN 2020024525 | ISBN 9789813146778 (hardcover) | ISBN 9789813146785 (ebook) | ISBN 9789813146792 (ebook other)
Subjects: LCSH: Mathematics--Study and teaching--Islamic countries.
Classification: LCC QA14.I742 M38 2020 | DDC 510.71/01767--dc23
LC record available at https://lccn.loc.gov/2020024525

British Library Cataloguing-in-Publication Data
A catalogue record for this book is available from the British Library.

For any available supplementary material, please visit
https://www.worldscientific.com/worldscibooks/10.1142/10201#t=suppl

https://doi.org/10.1142/9789813146785_fmatter

Dedication

Mathematics and its Teaching in the Muslim World is dedicated to Dr. Ahmed Djebbar whose contribution to the history of mathematics in the world of Islam reaches far beyond the information included in this volume. Dr. Djebbar's *L'âge d'or des Sciences Arabes* is considered a contemporary classic in the field of Islamic science and mathematics. Dr. Djebbar is also the author of *Une Histoire des Sciences Arabes*, and *Une Histoire des la Science Arabe (Entretiens avec Jean Rosmorduc*) and scholarly articles all demonstrating deep knowledge of Islamic scholarship.

Dr. Ahmed Djebbar was born in Aïn Defla, a wilaya in northern Algeria. He is a mathematician, and historian of science and mathematics.

He worked at University of Lille and holds the rank of Professor Emeritus. Dr. Djebbar is also a researcher in the history of science at the Paul Painlevé Laboratory (CNRS), which specializes in mathematics from the Muslim West. He was also previously adviser of Algerian President Mohamed Boudiaf, until 1992. From July 1992 to April 1994 he held the post of Minister of Education and Research in Algeria in the governments of Belaid Abdessalam and Redha Malek.

https://doi.org/10.1142/9789813146785_fmatter

Contents

https://doi.org/10.1142/9789813146785_fmatter

Preface

The economic, political and cultural importance of the Muslim states of the Middle East has increased exponentially in the century from World War I to 2016. Their strategic location at the intersection of Europe, Asia and Africa and the petroleum reserves they share focused world attention upon them.

The role of the Muslim religion has shaped their common culture in ways unfamiliar to Europe, Asia and the Americas. These cultural differences have been especially important in education and even in fields like mathematics that appear to western scholars to be independent of spiritual concerns.

The vast petroleum wealth of many Middle Eastern nations has financed development in a number of areas including education, but with surprisingly little effect upon world scholarship. Mathematics from the UNESCO 1966 Project for the Improvement of Mathematics Teaching in the Arab States led to the expansion of America and European university campuses into the Middle East but seemed only to transpose western educational methods to Middle Eastern settings rather than to influence traditional Muslim educational values.

The purpose of this anthology is to review the history of mathematics teaching and learning from traditional Muslim schools to contemporary practice in the schools of thirteen Middle Eastern nations.

The nations examined range from large to small, from rich to economically struggling, from religiously conservative to open and tolerant. It is not surprising that these differences affect even the teaching and learning of mathematics. The diversity of the educational and cultural environments of the thirteen nations included in this anthology may lead to different interpretations and conclusions, despite the fact that contributors to the anthology are national experts in the field of mathematics education.

In this situation, when different interpretations occur, it is the responsibility of the editors to insure that interpretations and conclusions are fair and accurate. Hopefully *Mathematics and its Teaching in the Muslim World* meets this standard.

If the descriptions of national accomplishments described in chapter after chapter are to be contrasted with those of European, Asian, and North and South American nations, it is essential to know more about Arab and Islamic achievements in past centuries than is commonly known by western scholars.

The editors are honored to begin the anthology *Mathematics and its Teaching in the Muslim World* with an introduction by Dr. Ahmed Djebbar, former Algerian Minister of Education, a member of the Algerian Academy of Science and Technology and a world-renown historian of Islamic science and mathematics. Dr. Djebbar recognizes the decline in Islamic science and mathematics in the 18th and 19th centuries from its apogee in the 12th to 13th centuries. His introduction is a distillation of the achievements of Arab and Muslim mathematics from classical times to the present emphasizing the importance of their achievements to Western mathematical accomplishments in subsequent centuries to the present.

His message to the readers of this anthology is to take pride in the rich history of Muslim mathematics by supporting and facilitating the work of Muslim leadership in study of historic Arab contributions including combinatorics, number theory and algorithms. Dr. Djebbar is not silent about mathematics education but rather emphasizes the need to pursue mathematics teaching and learning with a passion evident in the work of North African and Middle Eastern scientists of the first two Millennia.

In our roles as advisors to literally hundreds of doctoral students, the editors see a wealth of research topics of interest to Arabic speaking students. For example, early contributions to combinatorics precedes the work of Pascal and the perennial Arab topic of fair division of an inheritance anticipates Arrow's work in the 20th century. Dr. Djebbar's introduction to this anthology should be "required reading" for mathematics graduate students world-wide.

Bruce R. Vogeli
Clifford Brewster Upton Professor and Director
The Program in Mathematics
Teachers College Columbia University
in the city of New York, NY

https://doi.org/10.1142/9789813146785_fmatter

Acknowledgements

The preparation of this anthology *Mathematics and its Teaching in the Muslim World* could not have been possible without the collaboration of many individuals with special knowledge of Islam and its languages and history. Manuscripts for individual chapters were received in English, Arabic, French, and Indonesian all of which required translation and interpretation into contemporary English.

In some cases, even English language manuscripts required editing. The anthology's editors Mohamed El Tom and Bruce Vogeli possessed some of the knowledge and skills required but relied upon assistance by editors and educational specialists including many chapter authors who went beyond their chapter responsibilities to advise and aid the anthology's editors.

In addition to these resources, other education and language specialists were generous in providing pro bono assistance.

Special thanks are also owed to the book's editorial staff:

Ms. Anne Renaud, Dr. Vogeli's capable corresponding editor and a PhD student of Teachers College, Columbia University's program in Clinical Psychology, was an essential contributor to the anthology during Dr. Vogeli's extensive hospitalization during the final months of the anthology's preparation.

Ms. Sonja Hubbert, an experienced compositor who worked on previous volumes in the series reviewing world mathematics education, provided the rapid and accurate services to maintain the anthology's ambitious production schedule in spite of one co-editor's illness.

Ms. Amanda Alexander, a research assistant who worked with Dr. Vogeli on previous projects returned to the Muslim World anthology to maintain both the project's schedule and quality. Ms. Alexander is a recent graduate of the Art & Art Education program and was responsible for design elements of importance to *Mathematics and its Teaching in the Muslim World.*

Dr. Michael Kent who attended a French institution, provided English translation of French manuscripts. Dr. Kent is a mathematics professor at a New York College. He is a graduate of Teachers College, Columbia University. The co-editors and publisher express their gratitude to these and other specialists, without whom this anthology could not have been completed.

Mohamed El Tom
Bruce R. Vogeli
Editors

https://doi.org/10.1142/9789813146785_fmatter

INTRODUCTION: Mathematics Activities in Islamic Countries and their Dissemination in the VIIIth–XVth Centuries

Dr. Ahmed Djebbar
Professor Emeritus at the University of Lille, France
Member of the Algerian Academy of Science and Technology

Between the eighth and fifteenth centuries important mathematical developments can be observed in the scientific centers of Islamic countries. Today, dozens of bio-bibliographical works and thousands of manuscripts scattered in public and private libraries throughout the world still attest to this.[1] These developments are the result of multifaceted practices that were thought, taught, written and published in Arabic, in one of the many scientific centers of Islamic countries. By extension, it also includes the mathematical developments expressed from the late eleventh century in Persian, Hebrew and sometimes even in Latin, but identified by historians of science as having the characteristics of the Arabic tradition in their components, terminology, and methods.

The various activities in this area were preceded and accompanied by intense scientific exchange, direct or indirect, that took place inside and outside the borders of the Muslim empire, between people and communities with different languages, religions, and cultures. These exchanges occurred through three main phases which overlapped in time and which are closely linked. The first took place mainly in Syria, Iraq, and Alexandria. It started in the late eighth century and ended around the mid-tenth century. This period witnessed the birth and development of a mass transfer phenomenon, through translations but also by other routes (oral communication, professional activities), of Greek, Indian, Babylonian and Persian knowledge. The results of this transfer were first expressed in Syriac (especially for medical texts), then exclusively in Arabic.[2]

The second phase, the longest, stretched from the early ninth century to the late twelfth and for some scientific centers of Central Asia until the mid-fifteenth century. This phase also led to many contacts and exchanges within the context of the scientific activities that developed.

The third and final phase took place mainly in the Muslim West (Andalus, Sicily, Maghreb); from the twelfth century; and involved the elites of medieval Europe. This phase corresponded with the period of direct distribution and then translation (in Latin, Hebrew, Castilian, and other languages) of a number of Arab scientific works or of Greek works available in Arabic versions. Just like the first phase, but on a larger scale, this period enabled two distinct cultural worlds that coexisted through trade and competed by wars (Eastern Crusades, Reconquista in the Iberian Peninsula and Sicily) to establish and develop new relationships based this time on scientific activities.

In these activities and in these exchanges, mathematicians played a significant role as producers of knowledge and as "service providers" for other disciplines. The content of their production during this long period of history has been the subject of much research. A part of this, quantitatively the most important, has been concerned with the evolution of ideas and mathematical methods. A second part concerned the links between this scientific work and its external environment. There are firstly the ancient heritages that have been the foundation on which a new mathematical tradition was formed. Then, with the widespread development of intellectual activities, strong links have been forged with other scientific fields (astronomy, physics) or broad cultural fields (linguistics, philosophy, astrology). Finally, from the twelfth century, the needs expressed outside the borders of the Muslim empire (especially in medieval Europe) opened new horizons to what had survived of Indian and Greek mathematics and their Arab-Muslim extensions.

We wish to make a few preliminary remarks that will help contextualize the contents of the Arabic mathematical tradition in relation to the preceding traditions and try to connect these new practices to the cultural, ideological, political and economic dimensions of their environment. The first remark concerns the geographical area in which Arab mathematical activities were born and developed. It spans three continents (Asia, Africa and Europe) and it brings together many populations which differ in language, culture, beliefs and know-how. Chronologically, only the center of the empire was concerned with the first scientific initiatives. To be more accurate, some cities in Syria, Iraq and Persia, Alexandria in Egypt must be added. But from the ninth century, new intellectual centers emerged at the periphery of the empire, particularly in Maghreb and Andalus.

The second remark concerns the specific nature of mathematical practices in Islamic countries. Firstly, we observe from the late eighth century a phenomenon of juxtaposition of collected knowledge with the beginning of adaptation to local conditions and needs. After assimilation of the essential part of the knowledge borrowed from earlier civilizations, the first scientists were concerned with ways to synthesize the various elements of this rich heritage, with its content of know-how, approaches and results. This synthesis made it possible to integrate into one scientific practice two major traditions that had developed, sometimes independently, over the centuries before the advent of Islam. The first, of Greek origin, is characterized by a hypothetical-deductive approach. It is a set of rules, methods, tools and expressions that help achieve a result and justify the validity of the demonstration. The second tradition includes all algorithmic processes that the first mathematicians of the Islamic countries discovered, directly, in local practice, that is to say, those of the Fertile Crescent societies,

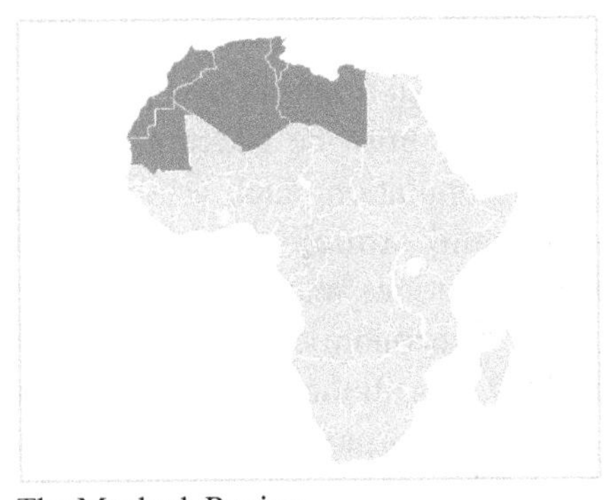

The Maghreb Region

or indirectly through what came to them from Indian, Chinese, Mesopotamian, or Egyptian sources.[3]

One final remark concerns the level of development of mathematics in different regions and populations of the empire. The information we have today confirms the pioneering role of Baghdad in the revival of science from the late eighth century. It also reveals that from the tenth century there was a standardization of the content and level of training, as a result of both rising education levels in different regions of the Muslim empire and the spread of written knowledge, expressed this time in a single language, Arabic. As a result of this phenomenon a number of scientific centers emerged and developed, distant from the capital of the empire.

Mathematics in the Muslim East

The Context of the Advent of a New Scientific Tradition

Among the key factors that were responsible for the birth of a new mathematical tradition, first in the center of the empire and then at its periphery, are the intellectual centers that existed before the rise of Muslim power and about which we have some information. Despite their limited number and the modest nature of their activities at the time of interest here, namely the seventh century and part of the eighth, these centers played an important role in starting science by enabling communities of different faiths (Muslim of course, but also Christian, Jewish, Sabean, Zoroastrian) to participate in their own right in the birth and development of a powerful mathematical tradition.

To this must be added all the know-how of certain categories of the urban population, such as surveyors, inheritance planners, craftsmen in wood and stone, architects, accountants of central and regional administrations and some categories of traders. In different ways, these actors in economic and social life spread, through their daily practices, and towards all the communities that coexisted then, a body of knowledge that was incorporated into the mathematical corpus being constructed in the same way as the scholarly knowledge in translations. This knowledge and expertise that was, in the past, reserved for members of certain communities or certain professions, would extend beyond the various community and professional groups and thus spread on a larger scale and with greater speed. This was made possible, it seems, by social mobility and the increased trade that was strongly favored by the new government and the new social strata. To this must be added a new, purely technological factor: the birth and development of the paper industry.[4]

Pre-Islamic Intellectual Centers

The most dynamic centers that existed on the eve of the Muslim conquest were Alexandria in Egypt, Antioch and Edessa in Anatolia, Kenesrin and Ra's al-Cayn in Syria and Gundishapour in Persia. Each in their own way, and according to the historical factors experienced, maintained intellectual activities in various fields such as philosophy, medicine, theology and, to a lesser extent, mathematics and astronomy.

Upon arrival in Egypt in 642, the first Muslim conquerors found in Alexandria a powerful Christian community whose elite had education centers, some men of science, and especially private libraries containing prolific books on philosophy, medicine, astronomy, and perhaps other subjects, such as mathematics. These books were honored during the Hellenistic period in Egypt, but are not mentioned by known Arab sources. Some of these libraries still existed in the ninth century with the great Christian translator, Hunayn ibn Ishaq (d. 873) having found Greek manuscripts there.[5] Through these institutions scientific activity was perpetuated, especially in medicine. No Alexandrian mathematician is reported for this period and, regarding astronomy, the last specialist in this city about whom we have some information is the famous John Philoponus (sixth c.). But it is not known if his small treatise on the astrolabe (the Greek version has reached us) was available in any library of Alexandria.[6] Indeed, the first Arab astronomers borrowed the principle of the astrolabe from Greek tradition. But they do not mention the sources that allowed them to make their first studies on this instrument. They are also silent on the possible origin of certain Alexandrian Greek manuscripts that enabled them to make the first translations of the major works in mathematics and astronomy.

In Persia, Gundishapour was an important scientific center at the time of the Emperor Khosro I (531–579). The city had benefited from the patronage of the Emperor and that of his son. It also benefited from the exodus of Greek scholars expelled by the Byzantine rulers, because of their philosophical activities or their public paganism. This exodus was particularly important in 529 when Emperor Justinian (483–565) ordered the closure of the Academy of Athens where philosophers and scientists were still working, such as Simplicius the famous commentator of Euclid (third c. BC) and Aristotle (d. 322 BC). But it is unclear if this Persian scientific center was very active at the time of the arrival of Arab horsemen in the area, and we do not know the role it played in the dissemination of the Persian astronomical and astrological works that were translated into Arabic. It is also not known if this city was a relay in the distribution of part of the Indian mathematical knowledge whose presence is attested in the Fertile Crescent in the second half of the seventh century and then in Baghdad in the late eighth.

The third major pre-Islamic tradition is that of the different Christian communities of Syriac expression that were scattered over Syria and Anatolia, whose populations had become subjects of the new power after part of Byzantine territory was taken by the Muslim conquerors. Unlike the community of Alexandria, these communities have left us clear evidence of the nature of their scientific activities in the sixth to eighth centuries and of who the actors were. Of these, some were interested in mathematics and astronomy, evidenced by a fragment of an anonymous Syriac translation of Euclid's "Elements" and a treatise on the astrolabe by Severus Sebokht (d. 667). He also knew Indian calculation methods and perhaps contributed to their spread in the new context created by Islamic conquests made in the name of the new religion.[7]

The Appropriation Period of Pre-Islamic Mathematical Knowledge

During the period extending between the last third of the eighth century through the middle of the tenth, the elites of the Christian communities of the empire's center played an important role. Those among the scholars who were dedicated to translations responded to requests from princes, merchants and even caliphs who became the first patrons of science. Later, with the spread of education and development of science, they were approached by people of different faiths who needed Arabic translations of interesting books for their training, development or research activity. In his book *The Sources of Information on the Categories of Scholars*, Ibn Abi Usaybi'a (d. 1269) identified 35 translators, the majority of whom, given their names, appear to be Christian. He also gave us the list of eleven patrons who financed the translations. Most of them are Muslims but there are also non-Muslims, such as Bishop Theodore and Ibn Qutrub.[8]

At this stage, it should be noted that the translation phenomenon is not exclusively concerned with scientific literature. Indeed, in this area, there was, from the early seventh century, an interest in all writings concerning military engineering, the techniques of astronomical astrology and of chemistry.[9] Secondly, writings on scholarly medicine were in demand. And it was not until the advent of the Abbasid dynasty and especially the rise of al-Mansūr (754–775) to the throne of the caliphate that we observe the first initiatives for the translation of mathematical and astronomical works.

During the first phase of the Abbasid Caliphate (late eighth to mid-tenth c.), this translation phenomenon continued and diversified. In addition to the medical books that the caliph al-Mansūr (754–775) had translated by Jurjus Ibn Jibril al-Batrīq,[10] he also financed the translation by Muhammad al-Fazari, of an Indian astronomical work bearing the generic title of *Siddhanta* (*Sindhind* in Arabic).[11] His son al-Mahdī (775–785) continued this policy of patronage in favor of science. But it is with his grandson, al-Hārūn Rashīd (786–809) that this phenomenon grew and became better organized, especially with the founding of the "House of Wisdom" in Baghdad. This was the first state institution in Islam which brought together scholars of different faiths or opinions from various fields such as philosophy, astronomy, mathematics, theology, and sometimes translating these works. We know, for example, that qu'an Nawbakht (d. about 777) translated writings on astronomy and philosophy in Persian for the Caliph. We also know that Ibn Abi Mansur (d. 830), who was the chief astronomer affiliated with this institution, devoted large sums to the translation of writings on medicine, astronomy and music. The brightest translators of his time, like Hunayn Ibn Ishaq, Ishaq his son (d. 910) and his nephew Hubaysh worked for him.[12] It should be noted that, outside this institution, other scientists also funded translations. This is the case of the philosopher al-Kindī (d. 850) and the brothers Banū Mūsā (ninth c.), three brilliant and wealthy mathematicians who had translated, among other works, *The Conics* of Apollonius (third c. BC) by Ibn Abi Hilal al-Himsi (d. about 883) and Thabit Ibn Qurra (d. 901).[13]

Other ancient mathematics disciplines interested translators or their sponsors as well. Number theory of Neopythagorean tradition is represented by two translations of *An Introduction to Arithmetic* by Nicomachus (second c.).[14] Arithmetic chapters of Euclid's *Elements* (Books VII, VIII and IX) were translated into Arabic several times, along with the other ten chapters of the book. For the important treatise *Arithmetic* by Diophantus it was not until the late ninth century or the beginning of the tenth that a part of the book was discovered. It would be translated by Qusta Ibn Luqa (d. 910).[15]

In geometry, the three most important themes of the Greek tradition were partially known by way of an accidental discovery of the manuscripts. The first deals with plane and solid geometry, in which the objects (lines and figures) are constructible with ruler and compass. The source is essential and consists of nonarithmetic chapters of Euclid's *Elements*. They were translated at least three times between the late eighth century and the beginning of the ninth. The second topic concerns the geometry of measurement, with the translation of two works of Archimedes, *The Measurement of the Circle* and *The Sphere and the Cylinder*. The other aspect of this geometry, namely, surveying and carvings, was also disseminated. But as it was not scholarly mathematics, bibliographers do not mention translations of manuals dealing with this theme. This could mean that in this area dissemination happened orally, in connection with certain corporate activities. The third topic of geometry includes various Greek contributions to the study of the properties of conic sections and their use in solving certain problems or optical astronomy. The basic text in this field is undoubtedly the *Book of Conics* by

Apollonius (third c. BC), to which must be added some texts like the one attributed to Archimedes on the construction of the regular heptagon.[16]

The Arab tradition of calculation borrowed from several sources of different scientific traditions. They found the alphabetic numeration to 27 digits (without the zero) in Greek astronomical writings. In writings from India or Persia, they borrowed both the decimal place value system of numeration with 9 digits (supplemented by zero) and the numerical algorithms relevant to this system, namely, addition, subtraction, multiplication, division and an extraction technique to find the square root of an integer or a fraction. With these inputs, a set of tools and procedures appears to have been used in the Fertile Crescent before the advent of Islam: digital numbering with a certain number of algorithms to enable, mentally, certain operations and sometimes even to solve problems.

The Production and Innovation Phase in Mathematics

Before discussing the content of Arabic mathematical production, it seems useful to note that the period from the ninth century to the late thirteenth century, which was very fruitful in terms of output and new scientific directions, was also marked, in a less visible way, by the cultural diversity of the different communities that made up the society of the time. This diversity assumes constant exchanges that crossed cultural boundaries. Apart from a few moments of tension or violence caused by specific economic situations, or by exceptional behavior of some local authorities or officials,[17] cultural diversity has been a factor of stability on the political front and an inspiration in the sciences in general and particularly in mathematics.

Returning to the guidelines that characterize the mathematical activities of the Islamic city, we see that from a partial acceptance of the ancient heritages, mainly Indian and Greek in the case of written sources, the new guidelines depended on the problems encountered in the study of Greek and Indian corpuses, on the questions still open or imperfectly resolved in earlier traditions and on the attractions of other scientific disciplines such as astronomy, physics and linguistics. This is what we will show through the contents of the various mathematical disciplines that have been the subject of study, teaching, research and publication.

The Science of Computation

Before the advent of Islam and during the period prior to the seizure of power by the Abbasids in 754, the inhabitants of different regions of the empire used different number systems (written, mental or instrumental) and computational algorithms (addition, multiplication, division, subtraction, square root). It was a skill transmitted orally or by direct initiation. Then, from the ninth century, scholarly mathematics spread thanks to translations. They then proceeded to intensively use alphabetic numeration (in astronomical activities) and the decimal system in other computational practices. Moreover, we observe that the development of scholarly mathematics from the ninth century did not marginalize the old calculation methods. They were recovered and, to distinguish them from the new knowledge derived from translations, they were regrouped under the name "open computing."

We have no evidence allowing us to reliably determine all sources of this expertise. One source is made up of remnants of the scholarly knowledge produced by the pre-Islamic scientific traditions of the region, i.e., those of ancient Egypt, Mesopotamia and Greece. This, it seems, is the case of conventional operations of arithmetic, the process of the inverse[18] and of the method of false positions, both used for finding the unknown of a linear type of problem.[19]

The publication of the *Book on Indian Computations* by al-Khwārizmī created a new opportunity for the decimal system. In a context of scientific ferment and computing initiatives, this tool became the source of a new tradition of computing to exist alongside the old methods before being adopted permanently through teaching and the dissemination of a number of textbooks. This tradition will be distinguished by the different names given to it (Indian computation, the science of dust figures, computing using the tablet) and by its manuals. The latter will adopt a similar structure (definition of the decimal system, presentation of the four arithmetical operations, beginning with addition, presentation process of the method for extracting the square root), with the addition of new chapters corresponding to local traditions (the rule of three, proportional distributions, the method of double false position and sometimes even a chapter on algebraic methods).[20]

The Science of Numbers

In the field of arithmetic (in the Greek sense), Baghdad mathematicians were interested in prime numbers whose first properties they discovered in Books VII, VIII and IX of Euclid's *Elements*. The first study, by Thabit Ibn Qurra (d. 901) concerned the amicable numbers and how to get them.[21] His Epistle was the first of a series of works that continued in the East until the thirteenth century.[22] Among them, research in recent decades has revealed the contribution of Ibn al-Haytham (d. about 1041) to an aspect of the "Wilson's Theorem" and the contribution of al-Farisi (d. 1321) on the decomposition of a number.[23]

Then, with the partial discovery of *Arithmetic* by Diophantus (about 250), a second tradition took shape. It started with studies of the content of the Greek work, like those of Abu l-Wafa' and Ibn al-Haytham. But none of these contributions has come down to us. Then other avenues were explored. The first concerned the resolution of indeterminate equation systems. The contributions that have come down to us are those of Abu Kamil (d. 930) in his *Complete Book of Algebra* and the works of al-Karaji (d. 1029).[24]

The second enabled research on Pythagorean triples and on certain categories of Diophantine equations.[25] Among the writings that have addressed these issues, there are those mathematicians of the tenth century, such as Abu l-Jud, al-Khazin, as-Sjizi and those of Ibn al-Haytham (d. 1041).[26] The third orientation was inspired by the Neopythagorian heritage and concerned, first, the study of finite numerical series. In this context, results were obtained on the summation of arithmetic and geometric sequences and on their use in different types of problems: those of "Archimedean" geometry which deals with the calculation of areas and volumes of certain plane or solid figures[27] and those of algebra where these sequences are a pretext for constructing problems leading to equations of different degrees.[28] Secondly, there has been the study of figurate numbers[29] defined by Nicomachus in his *Introduction to Arithmetic*. Despite the lack of implementation of the results of this chapter, his teaching had some longevity and its contents circulated in various scientific centers of the Muslim empire.

Different Fields of Geometry

In geometry, a first contribution originated in the critical rereading of Euclid's *Elements*. This enabled the production of many books criticizing, commenting on or extending the content of the geometric chapters of the *Elements*, especially Books V and X. The in-depth study of their content enabled a gradual extension of the concept of number. Firstly, this allowed the manipulation of like numbers, of quadratic irrationals and every irrational obtained as an nth root of an integer or a fraction and finally, any ratio of two incommensurable magnitudes.[30]

Continuing this work, the mathematicians of Islamic countries were interested in all the problems that their Greek predecessors had not been able to solve using the tools of Euclidean geometry. Among these problems, the most famous is that of the division of a sphere in a given ratio. It corresponds to proposition 4 in Book II of *The Sphere and the Cylinder* by Archimedes (d. 212 BC). But there is also the problem of the multisection of an angle, the problem of determining n mean proportional, between two known magnitudes and the construction of regular heptagons and nonagons.[31]

In addition to old problems, mathematicians of the Arab tradition have studied different well-known geometrical objects and described items. This is the case of various studies on conic sections by Ibn Qurra Thabit,[32] as-Sijzi[33] and Ibrahim Ibn Sinan. Especially the case of the study, by Ibn al-Haytham, certain curves of degree greater than two that failed us but that is explicitly evoked by 'Umar al-Khayyam (d. 1131). This is the use of new curves whose intersections provide positive solutions to the equations of the fifth degree.[34]

To stay in the field of “learned” geometry we must also mention the Arab Archimedean tradition. From a partial knowledge of the works of

Archimedes (of which only two books were found and translated into Arabic), the first researchers in the field attempted to determine the area of a number of plane figures and solids and the volume of solids they generate by rotation about an axis. First there was the work of the brothers Banu Musa (ninth c.), entitled *Book on the Determination of Surfaces of Flat and Spherical Figures*,[35] then those of Ibn Qurra Thabit on the parabolas, ellipses and dishes, and those of Ibn Ibrahim Sinan (d. 940).[36] They will be followed by contributions of al-Kuhi in the tenth century and those of Ibn al-Haytham in the first half of the eleventh, for the volume of the sphere and that of the spherical dish.[37]

A few words on applied geometry, a discipline whose content reveals the important role of mathematics as problem solving tools that have arisen in the development of other scientific disciplines, such as optics, theoretical astronomy and astronomical technology. In response to the needs of designers, the Abu l-Wafa' (d. 997) published an important book in which he outlines different methods of construction of geometric figures and, in particular, methods for cutting a figure into several other figures or to make a new figure from other smaller but similar figures.[38] To stay in the field of flat decoration, it is helpful to note the contribution of Ibn al-Haytham (indicated by decorators) to perform certain tilings of the plane.[39]

In *Optics*, the writings of al-Kindi (d. 873), Ibn Sahl (tenth c.) and Ibn al-Haytham often went further than those of their Greek predecessors, especially in the study of plane and solid figures that can perform, under certain conditions, the reflection or refraction of light rays.[40]

In architecture, the realization of ground planes and the design of some volumes needed geometric knowledge. Unfortunately, the documents confirming this fact are rare. One of the most important is the work of al-Kashi (d. 1429), titled *The Key Calculation*. Its author exposes methods for making domes, arches and decorations three-dimensional (muqarnas).[41]

Finally, there is the geometry of astronomical instruments and, in particular, the different types of astrolabes whose implementation began in the late eighth century. The most important books on this subject are those of al-Biruni (d. 1048) and al-Murrakushi (thirteenth c.).[42]

Alongside all these technical activities, a number of studies have involved three important aspects related to the foundations of hypothetical-deductive geometry and tools that helped develop it: axiomatic postulating parallels,[43] problematic definitions, and demonstration processes.[44]

The Art of Algebra

In the early days, algebraic practices were considered separate from the science of calculation and were characterized as "art" as opposed to "science." In any case, that is what we deduce from the writings of al-Khwārizmī (d. 850), the author of what is considered today to be the first treatise on algebra. In fact, in the introduction, he describes its contents as a "précis of calculation by restoration and comparison."

The book is in two parts: the first begins by rapidly describing the decimal system and then defining the objects of algebra, namely numbers (strictly positive), the unknown, the square of the unknown and six equations it is possible to combine using these three elements. After that, the author outlines equation solving methods, followed by demonstrations to validate the existence of their positive solutions. Al-Khwārizmī concludes the first part of his book by presenting the tools to "mathematize" a problem and express it as an equation.

The second part of the book is designed as a set of chapters applying the tools of the first part. The author shows, through examples, how to solve problems of surveying, commercial transactions and division of inheritances.[45]

Bibliographers and mathematicians who refer to this book imply that other books on algebra were published at the same time. Part of one of them, attributed to Ibn Turk (ninth c.), is known to us. Its content is not fundamentally different from that of the book by Al-Khwārizmī[46] which from the tenth century became the standard reference.

Subsequent algebraic sources after the ninth century enable us to track the progress of the discipline on a technical and theoretical level. Here we see its independence in relation to the science of calculation and geometry. These sources also show the extension of its scope and identify its various interventions in other disciplines, such as astronomy, optics and even the science of inheritances.

The first progress is visible in the work of Abu Kamil who extended arithmetic operations to the objects of algebra (unknowns and monomials) and who used irrational numbers with great ease in solving equations (both as coefficients and roots). He said further research was conducted and new directions emerged: the development of the algebra of polynomials by the successors of Abu Kamil as Sinan Ibn al-Fath (tenth c.), al-Karaji and al-Samaw'al (d. 1175), unspecific analysis by the same al-Karaji, theory of cubic equations by Abu l-Jud, al-Khayyam and Sharaf ad-Din al-Tusi (d. 1213), approximation methods by Ibn Labban (tenth c.), al-Tusi and later by al-Kashi (d. 1429).

Trigonometry Used in Astronomy

In the Arab scientific tradition, trigonometry is a continuation of a legacy that is both Greek and Indian. It developed mainly in the context of astronomical activities; but its objects, tools and procedures have served other disciplines. Among the astronomical activities that stimulated the development of this new chapter was the study of planetary movement, calculating time, the design of instruments and, above all, the making of hundreds of tables for multiple and varied uses. These tables, regularly updated, provided a wealth of information about the calendars of the various communities in the Muslim empire, the values of latitudes and longitudes of many cities, the direction of Mecca for hundreds of cities in the Muslim empire, the positions of the seven visible planets, the moments of conjunctions between some of them, etc.[47]

From the end of the eighth century, some Baghdad astronomers preferred Indian trigonometric concepts to those of Ptolemy. They borrowed from them, in particular, the small table of sines and increased its volume throughout the ninth century. Then they introduced new trigonometric lines like the tangent, cotangent, secant and cosecant.[48] In the first half of the tenth century, they established the fundamental relationships between the six trigonometric magnitudes and they started to look for new trigonometric relationships. They thus established new results, the most important of which to them was the “theorem which provides,” that is to say, the spherical version of the famous theorem of the sines.[49] Among the researchers who contributed to this progress were Abu l-Wafa', Ibn 'Iraq (eleventh c.) and al-Biruni.

The last phase of the History of Arabic trigonometry was its progressive autonomy. First there was the appearance in astronomical treatises of separate chapters devoted to its objects and tools of this discipline, as evidenced by the content of the *Epistle on the Arcs of the Sphere* by Ibn 'Iraq or of the *Book of the Almagest* by Abu l-Wafa'. Then, authors devoted books exclusively to the tools and results of trigonometry. This is what Nasir al-Din at-Tusi does in his *Book of the Secant Figure*.[50]

Combinatorial Analysis

From the late eighth century, some combinatorial practices emerge in the East. They are associated with concerns related to the Arab metric and lexicography and al-Khalīl Ibn Ahmad was the initiator. Then mathematicians took over within the context of certain astronomical or algebraic operations. Enumerations and elementary numbering appear in an epistle by Ibn Qurra Thabit on the composition of connections and in a treatise by al-Biruni entitiled *The Keys to Astronomy*. In algebra, we find similar processes used by Abu Kamil who studies in his *Book of Rare Things in Calculation*, the resolution of some systems of indeterminate equations set out in the form of bird problems. Al-Samaw'al (d. 1173), enumerated equations or solutions to a given problem.

Two other contributions are noteworthy because they provide information about the nature of the problems associated with combinatorial practices and persistent combinatorial concerns in the tradition of Arab mathematics. The first is by al-Farisi (d. 1319), a scientist from Central Asia whose major contributions are in the field of optics. In his treatise entitled "Memorandum for Friends Explaining the Proof of Amicability" he is interested in the decomposition of an integer into prime numbers products. This leads him to build the arithmetical triangle, according to a combinatorial process, and use it to determine the number of divisors of a given integer.[51] A century later, the Egyptian mathematician Ibn al-Majdi (d. 1437) presents in his *Book of Substance*, a combinatorial method for determining the number of equations of any degree from a given number of monomials used to compose the equation.[52]

Magic Squares

Bibliographic and mathematic sources are silent about the origin of the chapter of arithmetic called “harmonious numbers” by specialists in Islamic countries and later called “magic squares” in medieval Europe. In any case, there was the publication, from the ninth century, of Arab writings on this subject. And the development of this chapter continued at least until the seventeenth century, with the spread of ancient construction methods and the discovery of new methods. Then, from the tenth century, and parallel to the purely scientific contributions, there appears a new astrological literature, with a therapeutic aim, based on the construction techniques of certain types of magic squares.

Based on the work undertaken recently on the history of this chapter, it seems that the first construction of magic squares in Islamic countries found their inspiration in the chess game that was practiced in Persia before the advent of Islam.[53] According to the first known treatise devoted exclusively to this topic, it is attributed to Thabit Ibn Qurra, the famous mathematician of the ninth century. This book has not reached us but it probably paved the way for other writers, such as al-Antaki (d. 987) and, especially, the Abu l-Wafa' (d. 997).[54] The analysis of the writings of these two authors show that in the tenth century the construction processes of “simple magic”[55] squares, of squares with borders,[56] squares with compartments[57] and some pan-diagonal squares[58] were mastered.

From the eleventh century general methods for building all the above kinds of magic squares[59] and new methods to streamline old methods appear. As these practices were considered part of number theory, it is not surprising that famous mathematicians were interested in this topic. This is the case, in particular, of Ibn al-Haytham (d. after 1040) whose work has not yet been found, but an abbreviated version has reached us.

According to the testimony of the Brethren of Purity, authors of the famous *Epistles* from the tenth century, astrologers were interested in magic squares and began to promote them by presenting them as talismans having certain virtues, especially in favor of painless childbirth.[60]

Greek, Arab and Indian Mathematics in the Western Mediterranean

The scientific activities that developed in the Muslim West, between the ninth century and the beginning of the fifteenth, were also expressed in Arabic. They started in Kairouan, then emerged in Cordova where their teaching began. It is assumed that at the time mathematics was taught to solve, in the first place, the problems associated with secular and religious practices of everyday life, especially related to commercial transactions and the division of inheritances. Then there was a phase of appropriation of mathematical knowledge from the East about which we do not have accurate and reliable information. It was not until the tenth century that the results of this long period of maturation appeared in the form of manuals of mathematical material from the East and in the form of original production. This scientific dynamic resulted in a rich flowering, first in Andalus between the middle of the tenth century and the middle of the twelfth, then in the Maghreb in the thirteenth and fourteenth centuries. With the total disappearance of Muslim political power in Sicily and the Iberian Peninsula, the Maghreb remained the only region where the Muslim West practices in mathematics continued in Arabic. During the long period stretching from the late fifteenth century to the late eighteenth, there was, in the most vibrant cities in the region, a halt in research, a restriction of the content of teaching programs and a decline in the material taught in comparison to what was practiced during the previous phase. In the chapters that follow, we will discuss the movement of Arab mathematical knowledge around the western Mediterranean, describing the essential elements constituting the disciplines taught in this region and referring to what is known today of their dissemination in Europe from the late tenth century, first anonymously and then through translations from the twelfth to the fifteenth centuries. But before discussing the content of mathematics texts that circulated or were produced in the Muslim West and those that circulated in Europe, it would seem useful to give some information about the context of this movement.

The Context of the Movement of Arab Mathematics in Europe (Tenth-Fifteenth Centuries)

From the last third of the eleventh century, the beginning of the "Castilian Reconquista" sought to recover the Iberian territories ruled by Muslims authorities since 711. This offensive resulted in significant consequences at the cultural and scientific level. One such consequence materialized in the powerful phenomenon of transfer of knowledge to Europe. Indeed, we see, first in a hesitant way towards the end of the tenth century and with more vigor and continuity from the beginning of the twelfth, the development of a phase of appropriation for philosophy, the exact sciences and their applications. This was done through a vast movement of translations from Arabic into Latin and Hebrew (then into local languages). It reached its peak apogee in the twelfth century but continued until the mid-fifteenth. It is also noticeable that this phenomenon concerned almost exclusively the western Mediterranean, in the sense that scientific centers of the East, still very active in the early twelfth century do not seem to have played a direct role.

We must emphasize that this is not a simple technical operation of translation or direct assimilation of the knowledge produced or preserved in the Muslim empire. It was instead a powerful intellectual movement, similar to that experienced by the East in the eighth and ninth centuries, which helped bring together hundreds of people from different countries, faiths and cultural backgrounds. In the western Mediterranean, these contacts were necessary to carry out the translation tasks. It is known that very few translators mastered Arabic as a mother tongue. Among these, there were Jews especially like Petrus Alfonsi, Ibn Dawud and Ibn Ezra. But there were also, fewer in number, Christians like John of Seville and Hugo of Santalla. The other translators had access to scientific and philosophical texts only through intermediaries mastering both Arabic and the local vernacular. This is the case of Michael Scot (d. about 1235), who worked in Sicily in the court of Frederick II (1220–1250). For his part,

Gerard of Cremona (d. about 1187) used the services of Galippus, a Mozarabe, for the translation of *The Almagest*. As for the famous Roger Bacon (d. 1292), he argued that most major translators of the twelfth century mastered neither Arabic nor Latin and Michael Scot himself was actually a pseudonym for translations done by an Andalusian Jew.[61]

What is even more interesting for our purposes is the existence of veritable multi-faith teams of translators. We know that Peter the Venerable brought into the group of Robert Ketton and Hermann of Carinthia, an Arabist named Muhammad. This was also the case in the thirteenth century for the team funded by the King of Castile Alfonso X (1252–1284).[62]

The Content of Mathematics Available in The Muslim West and What Circulated in Europe

The Science of Computation

The oldest work on computation in the Muslim West, written in Arabic, was published in Kairouan by Abu Sahl al-Qayrawani in the ninth century. No copy has survived but its title, *Book of Indian Computation*, clearly indicates the Eastern origin of its content and its direct affiliation with the work of al-Khwārizmī. For the tenth and eleventh centuries, some names of mathematical practitioners are mentioned by bibliographers but no work produced in the region is mentioned.

The oldest computation manuals that have been preserved were published in the second half of the twelfth century by three authors, al-Hassar (twelfth c.), Ibn Muncim (d. 1228) and Ibn al-Yasamin (d. 1204). The first two are from al-Andalus and taught for some time in Marrakech. The third is from outer Maghreb but he lived and taught for several years in Seville. Their works are direct evidence of two closely linked phenomena. The first is the dissemination, on a greater scale than before, of part of the Andalusian tradition of the science of computation to scientific centers of the Maghreb, like Ceuta, Fez, Marrakech, Bejaïa and Tunis. The second phenomenon is the revitalization of teaching and research activities in the Maghreb, particularly thanks to the patronage of the first Almohad caliphs and then of dynasties that shared their empire.

The first mathematician left us two books on the science of computation. The smallest, called *Book of Demonstration and Recall* is a manual on numeration, arithmetic operations on integers and fractions, extraction of the exact or approximate square root of a whole or fractional number and the summation of series of integers (natural, even or odd) of their squares and their cubes. This book was circulated in Europe in a Hebrew translation made in 1271 by Moses Ibn Tibbon (d. 1340). The second work by al-Hassar is entitled *The Complete Book on the Art of the Number*. Only the first part of this work has survived. Its content expands on the topics of the part of the first book on integers and includes new chapters on the decomposition of a number into prime factors, dividers and common multiples on the extraction of the cube root of an integer. The second part of the work, of which we have only the table of contents, deals with operations on fractions, the summation of different categories of integers and the demonstrations of algorithms to calculate perfect, deficient, abundant and amicable numbers.[63]

The second author is relatively better known than the previous two. According to his biographers, his mother, whose first name was Yasamin [Jasmine Flower], was black (color he inherited) and his father was from the Berber tribe of Banu al-Hajjaj. For a long time, this mathematician was known only by his "Poem on Algebra." But we know today that he wrote a much larger work entitled *Book on the Fertilization of Minds with Symbols of Dust Figures*. It presents the contents of the classic chapters of the science of computation with additions on practical geometry. The content of this book illustrates the transitional phase that the twelfth century was and during which were juxtaposed, before melding into the same mold, three mathematical practices: that of the East, of al-Andalus and of the Maghreb.

The third mathematician of this period was born in Dénia (on the east coast of the Iberian Peninsula, near Valencia), but spent much of his life in Marrakech. The content analysis of his book (only one copy has survived), shows that this is not always a simple repetition of techniques and mathematical results belonging to the tradition of Indian computation. There are also "theoretical" chapters such as the study of figured numbers, the determination of amicable numbers and the solving of combinatorial problems.

The works of the three authors we have just mentioned were, along with other writings that have disappeared, the basic material for teaching the science of computation in the fourteenth and fifteenth centuries. The most important follower of this tradition was Ibn al-Banna who published two books, *Summary of the Operations of Computation* and *Lifting the Veil on the Operations of Calculation*. The second is both a mathematical and philosophical commentary on the content of the first. But it was the *Précis* that was the most successful and was taught and commented, until the end of the fifteenth century by various authors from al-Andalus, the Maghreb and even the Muslim East.

The first elements of computation in the Arabic mathematical tradition to reach Europe were the 9 digits and zero of Indian numeration that al-Khwārizmī had popularized with the publication of his *Book on Indian Computation*. These digits called "Arabic" were present in Latin texts from the late tenth century.[64] They are found a little later in charts and this initiative is generally attributed to Gerbert of Aurillac (d. 1003).[65] But the oldest known manuals which contain a detailed account of this numeration system and the arithmetic operations that accompany it, were translations of al-Khwārizmī's manual (d. 850).[66] These supplied the content of education books from the twelfth to the fourteenth centuries. Some were written in Hebrew, like *The Book of Number* by Abraham Ibn Ezra (d. 1167).[67] Others, more numerous, were in Latin, like *De Algorismo* by Jean de Sacrobosco (d. about 1244),[68] *Carmen de Algorismo* by Alexandre de Villedieu (d. 1240)[69] and especially *Liber Abaci* by Fibonacci (d. after 1240).

Known sources do not mention other Arabic works on the science of computation of Oriental origin that would have benefited from a translation in Latin or Hebrew. But for the computational tradition of the Muslim West, two manuals interested translators although we do not know the extent of dissemination of their contents. The oldest, *Book of Demonstration and Recall*, was published by al-Hassar (twelfth c.). It was translated in the thirteenth century, in the south of France, by Moses Ibn

Tibbon (d. after 1283).[70] The second is *Summary of the Operations of Computation* by Ibn al-Banna (d. 1321) translated in the fourteenth century, in Palermo, by Ibn al-Ahdab.[71]

Algebra

Compared to what was produced as part of this discipline in the East, what actually reached the Muslim West and Europe was relatively small and we continue to question the reasons for this modest circulation.

In the present state of our knowledge, we know that the algebraic writings of al-Khwārizmī and Abu Kamil arrived in Andalus, probably before the eleventh century, and they were discussed there.[72] They were responsible for the birth and development of a local algebraic tradition from which some representative writings have survived. In Andalus, the historian Ibn Khaldūn (d. 1406) cites many commentaries made in the region on the *Complete Book* by Abu Kamil. The oldest book in this category which circulated in the Maghreb was by al-Qurashi (d. 1186), a mathematician from Seville. It inspired another book written in Marrakech by Ibn al-Banna (d. 1321), the *Book of the Foundations and Premises of Algebra*, which was studied in the Maghreb from the fourteenth century and was also circulated in Egypt.[73] A third book, part of the Eastern tradition of the ninth and tenth centuries, was published by a certain Ibn Badr but we have no information on it.[74]

Alongside these works entirely devoted to algebra, we must mention the manuals on various aspects of the science of calculation which contain one or more chapters on algebra. The oldest are the *Book of Fertilization of Minds on Operations using Dust Figures* by Ibn al-Yasamin (d. 1204), *Summary of the Operations of Computation* by Ibn al-Banna and *Lifting the Veil on the Operations of Calculation*, by the same author. This type of book continued to be favored by teachers and other manuals of the same kind were published in the fourteenth and fifteenth centuries. This is the case of *The Sucking of Nectar* by al-Qayrawani (fourteenth c.) and *The Removal of the Tunic*, by al-Qalasadi (d. 1486).

A third category of writings was written in the form of easy to remember poems. Their text served as cards for teachers who only had to comment on every line in front of their students. The oldest of these poems, devoted entirely to algebra, was published by Ibn al-Yasamin.[75]

It was in Toledo, in the twelfth century, that the treatises on algebra by al-Khwārizmī and Abu Kamil benefited from the first translations.[76] The dearth of Arab sources available suggests that more important treatises on algebra, by al-Karaji and al-Khayyām, did not reach Andalus. The same observation can be made about other chapters of algebra, such as Diophantine analysis and the theory of polynomials. Important publications dealing with these topics were published between the tenth and the late twelfth centuries by Abu l-Wafa' (d. 997), al-Karaji and al-Samaw'al. No trace of their content was found in known Andalusian writings or those published in the Maghreb, with the exception of a late treatise published in Tunis at the end of the fourteenth or early fifteenth centuries by an author of Egyptian origin. This observation allows us to say that these contributions could not have happened in Europe, unless we assume that they followed the route of the Crusaders. But studies of these Christian expeditions to the East, and their possible role in the circulation of knowledge, have not confirmed this hypothesis.[77]

Regarding the forms of dissemination of algebraic production in the Muslim world to the first scientific centers of medieval Europe, bibliographic sources reveal that the book by al-Khwārizmī was the object of a Hebrew adaptation[78] and three Latin translations (those of Gerard of Cremona (d. about 1187), Robert of Ketton (about 1141) and William of Lunis (thirteenth c.). The second Arab algebra book that received two translations, in Latin and in Hebrew, is the *Complete Book* of Abu Kamil. The Latin version is anonymous. The Hebrew version was written in Italy by Mordechai Finzi (d. 1475). Mention should also be made of more modest writings that had Latin versions. Some were devoted to the problems solved by algebra, such as the *Liber Mensurationum* by Abu Bakr, an as of yet unidentified mathematician, others were calculation manuals containing a chapter on algebra. This is the case of the *Book of Demonstration and Recall on the Science of Dust Problems*, by al-Hassar (twelfth c.), translated into Hebrew, in Montpellier in 1271 by Moses Ibn Tibbon (d. 1283) and the *Précis of Operations of Calculation* by Ibn Banna (d. 1321) which was translated also in Hebrew, in the early fifteenth century, by Isaac Ibn al-Ahdab, a Spanish astronomer who traveled to the Maghreb and lived for some time in Sicily.[79]

The second route by which algebraic knowledge circulated could be classified as direct, to the extent that it avoided the mediation of a translation. They were new books written in Latin or Hebrew whose content had been studied and assimilated in Arabic by their authors. Some of these writings were simply taken from the material already covered in previous Arab manuals before the twelfth century. Others contain contributions that have not yet been identified in writings in Islamic countries and that could be new contributions of the authors. In this field, representatives of the Jewish tradition are present. The oldest is Abraham Bar Hiyya (d. about 1145), a scientist from Zaragoza. In his *Book of the Surface and Measurement*, he solves measurement problems using algebraic procedures of the Babylonian tradition, without copying the process of al-Khwārizmī.[80] His book was translated into Latin in 1145 by Plato of Tivoli with the title *Liber Embadorum*. The second author is Abraham Ibn Ezra (about 1167), which also deals with measurement issues, according to the process of his predecessor, in his *Book of Numbers*.[81]

As for books written directly in Latin and containing chapters of Arabic inspired algebra, they have a broader spectrum of methods used and problems addressed. Two of them were analyzed. The oldest is *Liber Mahameleth* by an anonymous author of the twelfth century who spent time in Castile. This is an important book that fits, as the title indicates, within the Arab tradition of al-Andalus which dealt with the mathematics of transactions. The first part of the book studies decimal notation, arithmetic operations on integers and irrational (quadratic and biquadratic) numbers. The second part studies arithmetic or algebraic methods for solving transaction problems in a broad sense: purchase, sale, payment of employees, forage consumption, basin filling, wort cooking, spending on oil for lighting, etc. But it should be noted that, despite the wealth of content, the book doesn't seem to have circulated widely in Europe. At least this is what we may assume from the few references to the work in mathematical writings subsequent to the twelfth century.[82]

The second Latin book with material drawn in large part from mathematical writings of the Muslim West and East, is *Liber Abaci* by Leonardo Pisano, better known as the Fibonacci (d. after 1240). He would have had his first mathematics training in Bejaïa, in Central Maghreb, and would have perfected it during his business trips in the Muslim East, in Byzantium and in Europe. Reading his book clearly shows that he had access to the content of *The Compendium* by al-Khwārizmī and the *Complete Book* by Abu Kamil. He seemed to also know about the progress of algebra after these two authors. In the third part of the fifteenth chapter of his treatise, he describes the objects and basic tools of algebra and then the algorithms for solving equations of two degrees or fewer, with geometric demonstrations to show the existence of positive solutions to these equations. The second part of the chapter is devoted to solving common problems, drawing from the works of the two Muslim mathematicians we have just mentioned.

The Theory of Numbers

Despite the scarcity of sources, we can say today that the three orientations of the theory of numbers that developed in the East (Euclidean, Diophantine and Neo-Pythagorean) are found in later writings in Andalus and the Maghreb but in highly variable forms. Thus, the theoretical aspect of amicable numbers was reproduced in al-Mu'taman's work,[83] in the form of a summary of *The Treatise of Ibn Qurra*, while their computational aspect is explained in the writings of al-Hassar (twelfth c.) and of Ibn Muncim (d. 1228).[84] Of the tradition inaugurated by the book of Diophantus, there remains only a few problems treated in algebra books without mention of their origin.[85] The third orientation is present in the writings of al-Andalus and the Maghreb, first as a research subject, among mathematicians of the twelfth century and then as chapters in calculation manuals.[86]

To our knowledge, no Greek or Arabic book devoted to number theory benefited from a Latin version during the great period of translations. But elements of this discipline were able to circulate either because they had been integrated into the Arab calculation manuals or because they have been copied for astrological purposes. This is the case, in particular, of playful aspects of arithmetic that were grouped in a chapter entitled "Thought Numbers." It is also the case of divination techniques of *zayrija*[87] and especially of magic square construction processes. Among European writings that have preserved traces of this circulation, there is the famous book of Bachet Méziriac (d. 1638), *Pleasant and Delectable Problems Made by Numbers*.[88]

Geometry

In addition to two Arabic versions of Euclid's *Elements*, by Ishāq Ibn Hunayn, (revised by Ibn Qurra) and al-Hajjaj (eighth c.), at least one commentary on their content, by Ibn al-Haytham, reached Andalus and the Maghreb, according to the testimony of the mathematician Ibn Haydūr (d. 1413). The same author also tells us that oriental writings on conics also circulated.[89] For his part, the philosopher Ibn Bajja (d. 1138), discusses the work of the Andalusian surveyor Ibn Sayyid (eleventh c.) on new curves and their use in solving the problem of the multisection of an angle.[90] The accuracy of his evidence confirms not only that the mathematicians of al-Andalus were in possession of advanced Oriental geometric work, but they were also aware of all open problems which their colleagues in other regions of the empire had been unable to solve.[91]

For the Archimedean tradition, the discovery of the book by al-Mu'taman (d. 1085) and its detailed analysis allows us now to say that the Oriental work of the ninth–tenth centuries was well known in Zaragoza and even stimulated research on some unresolved issues.[92]

As for the tradition that focused on the tools of geometry, it seems that at least one book, *The Book of Analysis and Synthesis* by Ibn al-Haytham, was studied and used by important authors, like al-Mu'taman Ibn Sayyid and Ibn Muncim. The latter even tried to extend the use of this type of demonstration to arithmetic proposals that were usually demonstrated by the induction method.[93]

As can be seen, a considerable part of geometric writings in the East were available either in Toledo or in other cities of al-Andalus. But, according to bibliographical sources, only a small portion of these writings benefited from translation to Latin or Hebrew. A first category consists of Greek works translated into these two languages from one of their Arabic versions or from one of their drafts made by a Muslim author. This is the case of the two most important books by Euclid, *Elements* and *Data*, or Books V–VII of *Conics* by Apollonius, drafted by Mahmud al-Isfahani (twelfth c.). A second category includes Arabic books produced in the East or in Andalus, starting from the ninth century: in particular, the *Book on the Determination of Areas of Plane and Spherical Figures* by three brothers Banū Mūsā (ninth c.), the *Treatise on the Secants* by Thābit Ibn Qurra, the *Book on Ratios* by Ibn ad-Daya (ninth c.), the commentary an-Nayrizi (tenth c.) the *Elements* by Euclid, the *Book of the Division of Surfaces* by Muhammad Baghdādī, and a chapter in the *Book of Geometry* of Ibn al-Samh (d. 1037). We should add geometric writings of philosophers like the drafting of the *Elements* inserted into the *Book of Healing* by Ibn Sīnā and the *Epistle* by al-Fārābī (d. 950) on the premises of Books I and V of the *Elements* that circulated in a Hebrew version.[94]

Trigonometry

We are certain today that all or part of the contents of some books on trigonometry written in the East circulated to the Muslim West and to Europe. This was done in several ways. The first was the circulation of the writings themselves, like the *Book of Healing* by Ibn Sīnā that contains precisely in its astronomical part, trigonometric results established in its time. The second is the direct assimilation of the contents of trigonometric works by astronomers of al-Andalus, during their scientific missions in the East. This was probably the case of Ibn Mucādh (d. after 1050) who studied in Cairo under Ibn al-Haytham and, back in his hometown, Jaen, published his *Book of Unknown Arcs of the Sphere* which contains most of the trigonometric results achieved in the eleventh century.[95]

As for the dissemination of the Arabic trigonometric corpus to Europe, we have some pieces of information that confirm the phenomenon and explain it somewhat. The first vector of this circulation seems to have been practical since it concerns many astrolabes, planispheres and universals, and introduced some basics of trigonometry, like the tangent, used to solve real problems of measurement (height of a building depth of a well, etc.). Concerning the theoretical side of the discipline, there were translations that helped make it known. In addition to the book by Ibn Sīnā, we have mentioned, there was the important treatise by the Andalusian astronomer Ibn Jābir Aflah (twelfth c.), *The Revision of the Almagest*, which contains not only basic trigonometric tools, but also the theorem of sines and a theorem attributed to the author that came to be known as Geber's theorem.[96] The famous Régiomantanus (d. 1476) who is credited with the first European book on trigonometry, *De Triangulis Omnimodis* knew the contents of the treatise by Ibn Aflah and used them.[97] It even seems that the contents of other Arabic writings on aspects of this discipline circulated one way or another beginning in the twelfth century.[98]

Two Original Contributions from the Muslim West

Among the contributions that developed in Andalus and the Maghreb and were somewhat innovative were mathematical symbolism and combinatorial analysis. The sources that have survived and have been studied in recent decades have revealed combinatorial and symbolic practices that are not, as with the practices or disciplines that we just mentioned, the result of the circulation of knowledge between the Muslim East and West. It even seems that, for these two areas, the circulation was in the opposite direction, as shown by some Eastern writings subsequent to the fourteenth century.

Arithmetic and Algebraic Symbolism

It seems that the first elements of a symbolism for expressing objects and mathematical operations emerged in the environment of Seville mathematicians in the twelfth century. In arithmetic, this involves the introduction of the fraction bar and letters symbolizing arithmetic operations like addition, subtraction and the square or fourth root. There is also a set of symbols to express different types of fractions used in transactions and in the division of inheritances. In algebra, only the letters of the alphabet are used to designate the unknown and its different powers, equality, subtraction. The first known appearance of this symbolism is in the *Book of Fertilization of Minds* by Ibn al-Yasamin.[99] But it is from the fourteenth century that its use seems to become generalized in the manuals of computation, as shown by the works of Ibn Ghāzī (d. 1513), Ibn Qunfudh (d. 1407) and al-Qalasādī (d. 1486).[100]

We find the symbolism of the fractions in *Liber Abaci* by Fibonacci (d. after 1240), yet this was not the only means of its dissemination in Europe. As for algebraic symbolism, F. Woepcke showed through a Latin document that it had seen a beginning of its circulation with the substitution of Arabic letters by Latin letters. But this circulation did not last long and did not generate new initiatives in the Latin scientific context of the twelfth and thirteenth centuries.[101]

Combinatorial Analysis and Its Applications

Elements of this discipline are identified, for the first time, in a book written in the late twelfth century in Marrakech, the *Book of the Science of Computation* by Ibn Muncim (d. 1228). In the eleventh chapter of this book, the author describes the procedures and formulas for counting the words of any language. It is also at that time that he built the famous arithmetic triangle (long attributed incorrectly to Pascal and Cardan) and identified the value of each part of the triangle to the number of combinations of n objects chosen from p objects. It is also from the construction algorithm of the triangle that he derives the formula. Ibn Muncim continued his research and established formulas for the permutations, with or without repetition, of a set of letters as well as the recurrence relation producing the number of possible interpretations of a word of n letters given all the diacritics of a language. These results, as well as others on the arrangements and combinations with repetition, allowed him to prepare tables that provide, by induction, all the numbering required.[102]

This contribution, quite new in the mathematical field of the time, is closely related to other Arab activities but dealing, since the eighth century, with linguistics, metrics and lexicography. And it seems that it was the successive failures of his predecessors that led Ibn Muncim to address this problem and find a complete solution. But his contribution was not an end point of the research that began in the East in the late eighth century (with the first results of Ibn al-Khalīl Ahmad (d. 795)). It actually inaugurated a new direction that would foster the emergence of an additional chapter to those that constituted arithmetic already. Indeed, in the second half of the thirteenth century or early fourteenth another mathematician from the Maghreb, Ibn al-Banna, revisited some of these results by linking them explicitly to the Neopythagorean Arithmetic and adding the arithmetic formula that avoids the use of the arithmetic triangle:

$$C_n^p = \frac{n(n-1)...(n-p+1)}{p(p-1)...2.1}$$

As an extension of these contributions, we see, in the Maghreb writings that have survived from after the thirteenth century, a broadening of the application field of formulas, reasoning and combinatorial approaches.[103]

We have no information concerning possible circulation of the book by Ibn Muncim in Europe. We discovered, recently, that the *Treatise* of Ibn al-Banna, which we just mentioned, was part of the library of the Jewish mathematician Ibn al-Ahdab (fourteenth c.). But he does not mention the combinatorial results in the book. During the same century or several decades before, another Jewish mathematician Levi Ben Gerson (d. 1344) was aware of some of the mathematical production of the Maghreb. He even translated a computation manual, published in Seville in the twelfth century by al-Hassar. In his *Book of Computation*, we discover combinatorial results as completed as those in the tradition of the Maghreb and which have no known affiliation with sources of Hebrew mathematical tradition.[104] This brings into question any possible, even partial, circulation of certain combinatorial texts of the Maghreb.

Conclusion

At the close of this presentation of the essential aspects of Arab mathematics activities between the late eighth century and the middle of the fifteenth, it will be clear to the reader that some important issues have not been addressed. First, the history of mathematics education, with everything that can be attached to it (infrastructure, program, teacher training, etc.). In the current state of research on this, and given the scarcity of sources, it is not possible to present these topics satisfactorily. Generally, bibliographers rarely describe the scientific practices as they were performed in daily life. As for the mathematicians, they sometimes mention the names of their teachers and the books they studied, but generally they do not talk about where they were trained and where they taught. Among the exceptions to this trend are the men of science who wrote types of autobiographies. But these writings mention neither the training of their authors or the content of their activities as teachers. They restrict themselves, in fact, to mentioning their teachers, books they studied under their guidance and the qualifications they obtained.

The second point we have not addressed concerns the different types of use of the tools and mathematical results in the resolution of problems in the context of other disciplines (astronomy, optics, science of inheritances, etc.). Unlike the first point, it is the abundance of material that made us abandon the detailed description of the role of some scientific disciplines in the development of tools, techniques or mathematical methods. We did it in a very succinct manner for trigonometry and combinatorial analysis, because these are chapters whose birth and development are directly dependent on other activities.

The third point concerns the phenomenon of the slowdown then halt of research activities in mathematics as well as the reduction process of the subject taught, sometimes eliminating certain disciplines in training programs, such as number theory, the geometry of conics and geometric optics. This question is more difficult to address than the other two because it is inseparable from the broader factors that have occurred in the dynamics of "decline" of the civilization of Islam. Among these factors, some might be classified as "external." This is the case of the progressive loss of the Muslim monopoly on trade over a wide radius. This is also the case of military offensives against the territories controlled by Muslim authorities and their consequences: first the Eastern Crusades and the Christian reconquest of Sicily and the north of the Iberian Peninsula. These events occurred between the late eleventh century and the beginning of the thirteenth. Then came the Mongol "invasions" which continued until the end of the fourteenth century. To these external factors, we should add those closely related to the dynamics of Muslim authorities and their political and ideological confrontations.

But the enumeration of these supposed factors is not sufficient to describe their effects on scientific dynamics and to identify the mechanisms at work in the process of widespread decline in all sectors of science. The only answer we can find at present concerns the state of science and especially mathematics in different regions of the Muslim empire from the late fifteenth century to the early nineteenth. The comparison, even partial, of the content of mathematical knowledge taught in centers of science in these regions shows that the slowdown phenomenon of scientific dynamics and the qualitative decline in its production first happened in regions such as Andalus and Central Asia which experienced profound disruption economically and socially. But the scientific decline of these regions would benefit the cities of neighboring regions. Next, the process of decline would affect all active scientific centers and all disciplines. For mathematics, this is clear from the content of manuals produced in the Muslim East or West.

References & Notes

[1] B. A. Rosenfeld & E. Ihsanoğlu: *Mathematicians, Astronomers & Other Scholars of Islamic Civilisation an their works (7th–19th c.)*, Istanbul, I.R.C.I.C.A., 2003.

[2] D. Gutas: *Pensée grecque, culture arabe*, Paris, Aubier, 2005.

[3] A. Djebbar: *Pratiques savantes et savoirs traditionnels en pays d'Islam: l'exemple des sciences exactes*, Actes du Colloque International sur "*Science and Tradition: Roots and wings for Development*" (Bruxelles, 5–6 avril 2001), Bruxelles, Académie Royale des Sciences d'Outre-Mer & UNESCO, 2001, pp. 62–86.

[4] M. Levey: Mediaeval Arabic Bookmaking and its Relation to Early Chemistry and Pharmacology, *Transactions of the American Philosophical Society*, New series, vol. 52 (1962), pp. 5–79: J. Pederson: *The Arab Book*, Princeton, Princeton University Press, 1984.

[5] M. Meyerhof: New light on Honain ibn Ishaq and his period, *Isis* 8 (1926), pp. 685–724.

[6] J. Philopon: *Traité de l'astrolabe*, J. Segond (ed. & trad.), Paris, Astrolabica, 1981.

[7] F. Sezgin: *Geschichte des Arabischen Schrifttums*, Leide, Brill, Vol. VI, 1978, pp. 111–112: F. Nau, Notes d'astronomie syrienne, *Journal Asiatique*, Série 10, t. 16 (1910), pp. 209–228.

[8] Ibn Abi Usaybi'a: *'Uyun al-anba' fi tabaqat al-atibba'* [The sources of information on the categories of doctors], Beyrouth, Dar maktabat al-hayat, sans date, pp. 279–284.

[9] A. Djebbar: *Le phénomène de traduction et son rôle dans le développement des activités scientifiques en pays d'Islam*. Proceedings of the symposium on « *Les Ecoles savantes en Turquie, sciences, philosophie et arts au fil des siècles* » (Ankara, 24–29 Avril 1995), S. Önen & C. Proust (ed.), Istanbul, Editions Isis, 1996, pp. 93–112.

[10] M.-G. Balty-Guesdon: *Le Bayt al-hikma de Baghdad*, Mémoire de D.E.A., Paris, Université de Paris III, 1985–86, p. 29.

[11] Sa'id al-Andalusi: *Kitab tabaqat al-umam* [Livre des catégories des nations], L. Cheikho (ed.), Beyrouth, Imprimerie catholique, 1912, pp. 49–50.

[12] D. Jacquart & F. Micheau: *La médecine arabe et l'Occident médiéval*, Paris, Maisonneuve & Larose, 1990, pp. 26–27.

[13] F. Sezgin: *Geschichte des Arabischen Schrifttums*, op. cit., vol. V, 1974, pp. 136–138.

[14] Nicomaque de Gérase: *Introduction Arithmétique*, J. Bertier (trad.), Paris, Vrin, 1978.

[15] J. Sesiano (ed. & trad.): *The Arabic Text of Books IV to VII of Dophantus' Arithmetica in the translation of Qusta ibn Luqa*, New York, Springer Verlag, 1982: R. Rashed (édit & trad.), *Les Arithmétiques de Diophante,* Paris, Les Belles Lettres, 1984.

[16] J.-P. Hogendijk: Greek and Arabic constructions of the regular heptagon, *Archive for History of Exact Sciences* 30 (1984), pp. 197–330.

[17] As was the case, for example, in Cairo with the whimsical decisions of the Fatimid Caliph al-Hakim (996–1021), and in Grenada, with the assassination, in 1066, of Ibn an-Naghrilla, a minister of the King, and its tragic consequences for the Jewish community of the city.

[18] The method is to start from the last operation stated in the problem and go back to the initial data by performing inverse arithmetic operation.

[19] The method of false position is to take a random number and check if it is solution of the problem. If it is not, which is generally the case, we perform the same operation with a second number at random. If, again, there is no solution of the problem, then we introduce these two numbers in a known formula. The calculation result is the exact solution of the problem.

[20] Ibn al-Bannâ: *L'abrégé des opérations du calcul*, M. Souissi (ed.), Tunis, Publications de l'Université de Tunis, 1969.

[21] Two numbers m and n are called "amicables" when the sum of proper divisors of m is equal to the sum of proper divisors of n.

[22] F. Woepcke: Notice sur une théorie ajoutée par Thābit ben Korrah à l'arithmétique spéculative des Grecs. *Journal Asiatique*, série 4, vol. 20, 1852, pp. 420–429: R. Rashed: *Entre arithmétique et algèbre*, Paris, Les Belles lettres, 1984, pp. 259–268.

[23] R. Rashed: *Entre Arithmétique et Algèbre*, Paris, Les Belles Lettres, 1984, pp. 227–299.

[24] J. Sesiano: Les méthodes d'analyse indéterminée chez Abu Kamil, *Centaurus* 1977, vol. 21, no. 2, pp. 89–105: J. Sesiano: Le traitement des équations indéterminées dans le Badi' fi l-hisab d'Abu Bakr al-Karaji, *Archive for History of Exact Sciences*, vol. 17, no. 4 (1977), pp. 297–379: A. Anbouba: *L'algèbre al-Badi' d'al-Karaji*, Beyrouth,

Publication de l'Université Libanaise, 1964: R. Rashed: *Entre arithmétique et algèbre*, op. cit., pp. 196–225.

[25] The Pythagorean triples are integers; the sum of squares of two of them is equal to the square of the third.

[26] A. P. Youschkevitch: *Les mathématiques arabes*, Paris, Vrin, 1976, pp. 66–69.

[27] Theses series intervene in the framing of a surface (parabolic) or volume (sphere, paraboloid), by rectangular surfaces or volumes. A. P. Youschkevitch: *Les mathématiques arabes*, op. cit., pp. 124–129.

[28] As the famous problem of Ceuta exposed by el-Hassar before the lawyers of this city: find the number satisfying the equation: $1^3 + 3^3 + 5^3 + \ldots + x^3 = 1225$. H. Suter: Das Rechenbuch des Abu Zakariyya al-Hassar, *Bibliotheca Mathematica*, II, 1901, pp. 12–40.

[29] A. Djebbar: *Enseignement et Recherche mathématiques dans le Maghreb des XIIe–XIVe siècles*, Paris, Université Paris-Sud, Publications Mathématiques d'Orsay, 1980, no. 81-02. pp. 76–89: A. Djebbar: Figurate Numbers in the Mathematical Tradition of Andalus and the Maghrib, *Suhayl*, Barcelone, no. 1 (2000), pp. 57–70.

[30] Book X is devoted to "binomes" and "apotomes" which are magnitudes whose measurement is expressed today in these forms: $m+\sqrt{n}$, $m-\sqrt{n}$, $\sqrt{m}+\sqrt{n}$, $\sqrt{m}-\sqrt{n}$. From the middle of the ninth century algebraists, such as al-Mahani, have extended this study to these values: $m^{\frac{1}{2^p}} + m^{\frac{1}{2^q}}$, $m^{\frac{1}{2^p}} - m^{\frac{1}{2^q}}$, and to others, regardless of their geometrical supports. They have also introduced them alongside integers and rational, in the resolution of equations. Ms. Paris, B. n. F., no. 2457 et Ms. Tunis, B. N., no. 16167.

[31] A. Anbouba: Construction de l'heptagone régulier par les arabes au 4e siècle de l'Hégire, *Journal for the History of Arabic Science*, 1978, pp. 264–269: A. P. Youschkevitch: *Les mathématiques arabes*, op. cit., pp. 93–94.

[32] L. Karpova & B. Rosenfeld: The treatise of Thabit Ibn Qurra on section of cylinder and on its surface, Archives Internationales d'Histoire des Sciences, no. 94, 1974, pp. 66–72.

[33] F. Sezgin: *Geschichte des arabischen Schrifttums*, Band V, Leide, Brill, 1974, p. 331.

[34] A. Djebbar & R. Rashed: *L'œuvre algébrique d'al-Khayyam*, Alep, Université d'Alep, 1981, p. 66.

[35] H. Suter: Über die Geometrie der Sohne des Musa ben Shakir, *Bibliotheca Mathematica*, 3, F. 3, 1902, pp. 259–272.

[36] B. Rosenfeld: *Geometrical transformations in the medieval East*. In: XIIe Congrès International d'Histoire des Sciences, III, A, 1971, pp. 123–131.

[37] A.-P. Youschkevitch: *Les mathématiques arabes*, op. cit., pp. 123–131.

[38] l-Wafa': *Kitab ma yahtaju ilayhi as-sani' min a'mal al-handasa* [Book on geometric constructions necessary for the craftsman], S. A. Al-^cAli (ed.), Bagdad, 1979.

[39] J.-P. Hogendijk: Mathematics and geometric ornamentation in the medieval Islamic world, *European Mathematical Society, Newsletter* no. 86, December 2012, pp. 40–41.

[40] M. Nazif: *Al-Hasan Ibn al-Haytham buhuthuhu wa kushufuhu al-basariyya* [Al-Hasan ibn al-Haytham, his research and discoveries in optics], Le Caire, 1942–43: R. Rashed: Pioneer in Anaclastics, Ibn Sahl on Burning Mirrors and Lenses, *Isis*, 81 (1990), pp. 464–491.

[41] A. S. Damirdash & M. H. Al-Hafni: *Miftah al-hisab* [The key of Arithmetic], Le Caire, Dar al-kitab al-'arabi li l-tiba'a wa l-nashr, 1967, pp. 176–188.

[42] Al-Biruni: *Kitab fi isti'ab al-wujuh al-mumkina fi san'at al-asturlab* [The comprehensive book about the possibilities of realization of the astrolabe], Ms. Leiden, Or. 591/4e: Al-Hasan al-Murrakushi: *Jami' al-mabadi' wa l-ghayat*, [The collection of principles and goals], Ms. Paris, BnF, no. 2507–2508: J. J. Sédillot: *Traité des instruments astronomiques des Arabes*, Paris, Imprimerie Royale, 1834.

[43] K. Jaouiche: *La théorie des parallèles en Pays d'Islam*, Paris, Vrin, 1986.

[44] A. Djebbar: *Quelques remarques sur les rapports entre philosophie et mathématiques arabes*, Actes du Colloque de la Société Tunisienne de Philosophie (Hammamet, 1–2 Juin 1983), *Revue Tunisienne des Etudes Philosophiques*, no. 2 (1984), pp. 3–21.

[45] A. M. Musharrafa & M. Mursi Ahmad: *Kitab al-jabr li l-Khwarizmi* [The algebra book of al-Khwarizmi], Le Caire, Dar al-kitab al-'arabi li l-tiba'a wa l-nashr, 1968.

[46] A. Sayili: *Logical necessities in mixed equations by 'Abd al-Hamid Ibn Turk and the algebra of his time*, Ankara, Türk Tarih Kurumu Basimevi, 1985.

[47] E. S. Kennedy: Late medieval planetary theory, *Isis*, vol. 57, 3, no. 189 (1966), pp. 365–378: D.A. King: Ibn Yūnus very useful tables for reckoning time by the sun, *Archives for the History of Exact Sciences*, vol. 10, no. 3–5 (1973), pp. 342–394: A. P. Youschkevitch: *Les mathématiques arabes*, op. cit., pp. 141–150.

[48] M.-Th. Debarnot: *Trigonométrie*. In *Histoire des sciences arabes*, R. Rashed (ed.), Paris, Seuil, 1997, vol. 2, pp. 163–198: C.A. Nallino: *Al-Battani sive Albatenii opus astronomicum*, Milano, Publicazioni del Reale Osservatorio di Brera, 1899–1907: A. Djebbar: *La phase arabe de l'Histoire de la trigonométrie*, Actes du colloque "*Les instruments scientifiques dans le patrimoine: quelles mathématiques ?*" (Rouen, 6–8 avril 2001), Paris, Editions Ellipse, 2004, pp. 415–435.

[49] Cette relation a été surnommée ainsi parce que, en ne faisant intervenir que quatre grandeurs (alors que celle de Ménélaüs en utilise six), elle réduisait les temps de calcul pour la confection des tables astronomiques. M. Th. Debarnot: *al-Biruni, Kitab maqalid 'ilm al-hay'a, La trigonométrie sphérique chez les Arabes de l'Est à la fin du Xe siècle*, Damas, Institut Français de Damas, 1985.

[50] A. Pacha Caratheodory: *Traité du quadrilatère*, Constantinople, 1891: M. V. Villuendas: *La trigonometria europea en el siglo XI*, op. cit.

[51] R. Rashed: *Entre arithmétique et algèbre*, op. cit., pp. 274–297.

[52] A. Djebbar: *Enseignement et Recherche mathématiques dans le Maghreb des XIIIe–XIVe siècles*, op. cit., pp. 107–112.

[53] J. Sesiano: *Un traité médiéval sur les carrés magiques*, Lausanne, Presses Polytechniques et Universitaires Romandes, 1996, pp. 7–10.

[54] . Sesiano: Le traité d'Abu l-Wafa' sur les carrés magiques, *Zeitschrift für Geschichte der Arabisch-Islamischen Wissenschaften*, Band 12 (1998), pp. 121–244.

[55] A square with simple magic is where the first *n* integers are distributed into square boxes so that the sum of each row, each column and each of the two main diagonals are equal.

[56] A magic square border or enclosure is one for which the removal of the outer edge provides a simple magic square.

[57] A magic square with compartment is one that consists of square compartments that also possess the property of simple magic.

[58] In a Pandiagonal square, the pairs of lateral diagonal satisfy the same arithmetic property of rows, columns and diagonals.

[59] J. Sesiano: *Les carrés magiques dans les pays islamiques*, Lausanne, Presses Polytechniques et Universitaires Romandes, 2004.

[60] Ikhwan as-Safa': *Rasa'il Ikhwan as-Safa'* [Epistles of the Brethren of Purity], Beyrouth, Dar Sadir, 1957, pp. 109–113.

[61] Ch. Burnett: *Some Comments on the Translating of Works from Arabic into Latin in the Mid-Twelfth Century*. In J. Vuillemin (ed.): Miscellanea Mediavalia, Cologne, G.-Diem, 1985, pp. 161–167.

[62] J. Vernet: *Ce que la culture doit aux Arabes d'Espagne*, G. Martinez Gros (trad.), Paris, Sindbad, 1985. pp. 123–142.

[63] M. Aballagh & A. Djebbar: Découverte d'un écrit mathématique d'al-Hassar (XIIe s.): le Livre I du Kamil, *Historia Mathematica*, no. 14 (1987), pp. 147–158.

[64] Burnett, Charles, « Fibonacci's ''Method of the Indians'' », *Bolletino di Storia delle Scienze Matematiche*, 2003, 23/2, p. 87–97: Ms. Rabat, Bibliothèque Nationale, no. K 222, fol. 5r.

[65] R. Latouche: *Richer, Histoire de France (888–995)*, Tome II, Paris, 1937: Burnett, Charles, « The Translation Activity in Medieval Spain », Jayyusi, S.K. (éd.): *The Legacy of Muslim Spain*, New York, Brill, 1994, vol. II, p. 1040: P. Portet: « Les techniques du calcul élémentaire dans l'Occident médiéval: un choix de lectures », Coquery, N., Menant, F., Weber F. (édit): *Écrire, compter, mesurer: vers une histoire des rationalités pratiques*, Paris, Presses de l'ENS-Editions rue d'Ulm, 2006, p. 51–66.

[66] Allard, André, *Muhammad ibn Musa al-Khwarizmi, le calcul indien (algorismus)*, Paris, Blanchard-Namur, Société des Etudes Classiques, 1992: Folkerts, Menso, *Die älteste lateinische Shrift über das indische Rechnen nach al-Khwarizmi*, Munich, Verlag der Bayerischen Akademie der Wissenschaften, 1997.

[67] Silberberg, Moritz, *Sefer ha-Mispar. Das Buch der Zahl, ein hebräisch-arithmetisches Werk des Abraham ibn Ezra*, Frankfurt, 1895.

[68] Grant, Edward, *Arabic Numerals and Arithmetic Operations in the Most Popular Algorism of the Middle Ages John of Sacrobosco*, Grant, E. (éd.), *A source book in medieval science*, Cambridge Mass., Harvard University Press, 1974.

[69] De Villedieu, Alexandre, *Carmen de Algorismo*, Halliwell, J.O., Rara Mathematica, 1839, Réimpression, New York: G. Olms Verl., 1977.

[70] Suter, Heinrich, Das Rechenbuch des Abu Zakariya el-Hassar, *Bibliotheca Mathematica*, 1901, 3/2, p. 12–40.

[71] Wartenberg, Ilan, *The Epistle of the number by Isaac ben Salomon ben al-Ahdab (Sicily, 14th century), a Episode of Hebrew Algebra*, Thèse de Doctorat, Paris, Université Paris VII, 2007.

[72] Ibn Khaldun: *Kitab al-'ibar* [Book of sentences], Beyrouth, Dar al-kitab al-lubnani, 1983, vol. 2, p. 899.

[73] A. Djebbar: *Le livre d'algèbre d'Ibn al-Banna.* In A. Djebbar: *Mathématiques et mathématiciens du Maghreb médiéval (IXe–XVe siècles): Contribution à l'étude des activités scientifiques de l'Occident musulman,* Thèse de Doctorat, Université de Nantes-Université de Paris-Sud, 1990.

[74] J. A. Sanchez-Perez (ed. & trad.): *Compendio de Algebra de Abenbeder*, Madrid, Imprimerie Ibérique, 1916.

[75] M. Abdeljaouad: *Ibn al-Ha'im, Sharh al-urjuza al-yasaminiya* [Ibn al-Ha'im, Commentary on al-Yasaminiya poem], Tunis, Publications de l'Association Tunisienne des Sciences Mathématiques, 2003.

[76] M. Levey: *The Algebra of Abu Kamil, in a Commentary by Mordecaï Finzi*, Madison-Milwaukee and London, 1966: J. Sesiano: *La version latine médiévale de l'Algèbre d'Abu Kamil.* In M. Folkerts & J. P. Hogendijk (ed.): *Vestigia Mathematica*, Studies in medieval and early modern Mathematics in honour of H.L.L. Busard, Amsterdam-Atlanta, Editions Rodopi, 1993, pp. 315–452.

[77] A. Djebbar: *La production scientifique arabe, sa diffusion et sa réception au temps des croisades: l'exemple des mathématiques*, Actes du Colloque International sur "*Occident et Proche-Orient: Contacts scientifiques au temps des croisades*" (Louvain-la-Neuve, 24–25 mars 1997), Bruxelles, Editions Brepols, 2001, pp. 343–368.

[78] T. Lévy: A Newly-Discovered Partial Hebrew Version of al-Khwarizmi's Algebra, *Aleph* 2 (2002), pp. 225–234.

[79] T. Lévy: L'algèbre arabe dans les textes hébraïques (I). Un ouvrage inédit d'Isaac Ben Salomon al-Ahdab (XIVe siècle), *Arabic Sciences and Philosophy*, vol. 13 (2003), pp. 269–301: I. Wartenberg: *The Epistle of the Number by Isaac ben Solomon ben al-Ahdab (Sicily, 14th century), An Episode of Hebrew Algebra*, Thèse de Doctorat, Tel Aviv, Université de Tel Aviv, 2007.

[80] Abraam Bar Hiia: *Llibre de geometria*, M. M. Guttmann & J. M. Vallicrosa (ed. & trad.), Barcelone, Editorial Alpha, 1931.

[81] T. Lévy: *Note sur le traitement des fractions dans les premiers écrits mathématiques rédigés en hébreu (XIe et XIIe siècles)*, P. Benoit, K. Chemla & J. Ritter (ed.): *Histoire de fractions, fractions d'histoire*, Basel-Boston-Berlin, Birkhäuser, 1992, pp. 280–286.

[82] J. Sesiano: *Le Liber Mahameleth, un traité mathématique latin composé au XIIe siècle en Espagne*, Actes du 1e Colloque maghrébin sur l'histoire des mathématiques arabes (Alger, 1–3 Décembre 1986), Alger, Maison du Livre, 1988, pp. 69–98: J. Sesiano:

The Liber Mahameleth, A 12th-century mathematical treatise, New York, Springer Verlag, 2014: A. Vlasschaert: *Le Liber Mahameleth*, Stuttgart, Franz Steiner, 2010.

[83] A. Djebbar: Les livres arithmétiques des *Eléments* d'Euclide dans une rédaction du XIe siècle: le Kitab al-istikmal d'al-Mu'taman (m. 1085), Revue *Lull*, Saragosse, vol. 22, no. 45 (1999), pp. 589–653.

[84] Ibn Mun'im: *Fiqh al-hisab* [The science of calculation], Ms. Rabat B. G. 416 Q, pp. 319–321.

[85] Ibn al-Banna: *Kitab al-usul wa l-muqaddimat fi l-jabr wa l-muqabala* [Book of foundations and preliminary on algebra]. In A. Djebbar: *Mathématiques et Mathématiciens du Maghreb médiéval (IXe–XVIe siècles): Contribution à l'étude des activités scientifiques de l'Occident musulman*, Thèse de Doctorat, Université de Nantes, 1990.

[86] Ibn al-Banna: *Talkhîs a'mal al-hisab* [The abridged book on the operations of calculation], M. Souissi (ed. & trad.), Tunis, Publications de l'Université de Tunis, 1969.

[87] *Zayrija*: Table of divination using manipulations of numerical sequences and operations on what is now called congruence classes.

[88] Bachet de Méziriac: *Problèmes plaisants et délectables qui se font par les nombres*, Paris, Editions Blanchard, 1993.

[89] Ibn Haydur: *Tuhfat al-tullab* [Adornment of students], Ms. Vatican or. 1403, f. 138b.

[90] A. Djebbar: *Abu Bakr Ibn Bajja et les mathématiques de son temps*, in *Festchrift Jamal ed-Dine Alaoui, Etudes philosophiques et sociologiques dédiées à Jamal ed-Dine Alaoui*, Publications de l'Université de Fès, Département de Philosophie, Sociologie et Psychologie, no. spécial 14, Fès, Infoprint, 1998, pp. 5–26.

[91] A. Djebbar: *Deux mathématiciens peu connus de l'Espagne du XIe siècle: al-Mu'taman et Ibn Sayyid*. In M. Folkerts & J.P. Hogendijk (ed.): *Vestigia Mathematica, Studies in medieval and early modern mathematics in honour of H.L.L. Busard*, Amsterdam-Atlanta, Editions Rodopi, 1993, pp. 79–91.

[92] J.-P. Hogendijk: *Le roi géomètre al-Mu'taman Ibn Hūd et son livre de la perfection (Kitab al-istikmal)*, Actes du premier colloque maghrébin sur l'Histoire des mathématiques arabes (Alger, 1–3 Décembre 1986), Alger, La Maison des Livres, 1988, pp. 51–66: J. P. Hogendijk: The geometrical part of the *Istikmal* of Yusuf al-Mu'taman ibn Hud (11th century), An analytical table of contents, *Archives Internationales d'Histoire des sciences*, vol. 41, no. 127 (1991), pp. 207–281.

[93] Ibn Mun'im: *Fiqh al-hisab*, op. cit., pp. 255–297.

[94] G. Freudenthal: La philosophie de la géométrie d'al-Farabi, Son commentaire sur le début du I^e Livre et le début du Ve Livre des Eléments d'Euclide, *Jerusalem Studies in Arabic and Islam* (1988), pp. 105–219.

[95] J. Samso: *"Al-Biruni" in al-Andalus*. In J. Casurellas & J. Samso (ed.): *From Baghdad to Barcelona, Studies in the Islamic Exact Sciences in Honour of Prof. Juan Vernet*, Barcelona, Anuari de Filologia XIX - Instituto Millas Vallicrosa, 1996, pp. 583–612.

[96] Ibn Aflah: *Geberi filii Affla Hispalensis de astronomie*, Libri IX, P. Apianus (ed.), Nuremberg, 1534.

[97] J. Regiomontanus: *De Triangulis Omnimodis*, B. Hughes (ed. & trad.), Madison, University of Wisconsin press, 1967.

[98] N. G. Hairetdinova: On Spherical Trigonometry in the Medieval Near East and in Europe, *Historia Mathematica* 13 (1986), pp. 136–146.

[99] T. Zemouli: *Al-a'mal ar-riyadiya li Ibn al-Yasamin* [The mathematical work of Ibn al-Yasamin], Magister en histoire des mathématiques, Alger, E.N.S., 1993, p. 17.

[100] A. Djebbar: *Enseignement et recherche mathématiques dans le Maghreb des XIIIe–XIVe siècles*, op. cit., pp. 41–54: M. Abdeljaouad: *Le manuscrit de Djerba: une pratique des symboles algébriques maghrébins en pleine maturité*, Actes du 7e Colloque maghrébin sur l'Histoire des mathématiques arabes (Marrakech, 30 mai–1e juin 2002), Marrakech, E.N.S., Imprimerie El Wataniya, 2005, vol. 2, pp. 9–98.

[101] F. Woepcke: Notice sur des notations algébriques employées par les Arabes, *Journal Asiatique*, 5e série, vol. 4 (1854), pp. 373–374.

[102] A. Djebbar: *L'analyse combinatoire au Maghreb: l'exemple d'Ibn Mun'im (XIIe–XIIIe siècles)*, Paris, Université Paris-Sud, Publications Mathématiques d'Orsay, 1985, no. 85-01.

[103] A. Djebbar: *Enseignement et Recherche mathématiques dans le Maghreb des XIIIe–XIVe siècles*, op. cit., pp. 67–75: M. Aballagh: *Raf' al-hijab d'Ibn al-Banna*, Thèse de Doctorat, Université de Paris I-Pantheon-Sorbonne, 1988.

[104] G. Lange (ed.): *Sefer Maassei Choscheb, Die Praxis des Rechners*, Frankfurt, 1909.

About the Author

Dr. Ahmed Djebbar is a world-renowned mathematician and science historian. He earned his doctorate in mathematics at the University of Paris in 1972, and a second doctorate in mathematical history form Nantes University in 1990. He is the author of many books, most notably *A History of Arabic Science* (2001) and *Discoveries in Islamic Countries* (2009). Dr. Djebbar is Professor Emeritus of History of Mathematics at the University of Science and Technology Lille. Additionally, he serves on numerous commissions and editorial boards and remains active in the global research community. Dr. Djebbar is known for his emphasis on the role of the history of science, both in the education system and in society at large.

https://doi.org/10.1142/9789813146785_0001

Chapter I

AFGHANISTAN: Education and Mathematics in Afghanistan

James Kennis
City University of New York

Geography and Demographics[1]

The Islamic Republic of Afghanistan is a landlocked country located in South and Central Asia and was founded in 1747 when Ahmad Shah Durrani united the regional Pashtun tribes. It is the 41st largest country with its territory covering 652,230 sq. km. (slightly smaller than the state of Texas in the United States). It is bordered by China (northeast), Iran (west), Pakistan (east and south), Tajikistan (north), Turkmenistan (north), and Uzbekistan (north).

It is estimated that Afghanistan has a population of 32.5 million Afghans making it the 41st most populous country (just below Canada and Morocco). The country's 2004 constitution recognizes 14 ethnic groups: Pashtun, Tajik, Hazara, Uzbek, Baloch, Turkmen, Nuristani, Pamiri, Arab, Gujar, Brahui, Qizilbash, Aimaq, and Pahai. The official languages are Afghan Persian or Dari (50%; lingua franca) along with Pashtu (35%). Turkic languages—mostly Uzbek and Turkmen—comprise 11% usage in the country while the 30 other minor languages and dialects are used by about 4% of the population. The citizens of Afghanistan are 99.7% Muslim (85–90% Sunni; 10–15% Shia) and 0.3% other religious denominations.

[1] Source: CIA World Factbook Afghanistan September 2015

The largest urban area is in Kabul (Capital) with 4.64 million people living there. Afghanistan is listed as having the highest infant mortality rate in the world (107.15 deaths/1000 live births) and has among the lowest metric for life expectancy (population mean = 50.87 years). The country-wide literacy rate—aged over 15 years and can read and write—is 38.2% with 52% of males and 24.2% of females being labeled as literate.

The government of Afghanistan is an Islamic Republic that has administrative divisions in 34 provinces. Their independence day is August 19, 1919, which is the date when U.K. control over Afghan foreign affairs was removed. Their current government is composed of three branches: Executive, Legislative, and Judicial. In the Executive Branch, the President has the title of Chief of State and also Head of Government and there is a cabinet of 25 ministers. In the Legislative Branch, there is a House of Elders that consists of 102 seats and the Judicial Branch has a Supreme Court and many, specialized subordinate courts.

Afghanistan's economy is beginning to recover from the 30+ years of continual conflict; mostly due to international economic assistance, but there is growth in the agricultural and service arenas. The international community is committed to serve Afghanistan's development. Security issues, weak government, corruption and lack of infrastructure are hampering this development, but progress is being made. In 2014, the GDP growth rate was 1.5% and the GDP per capita was $1,900. The labor force is mostly agriculture (78.6%), industry (5.7%), and services (15.7%). Opium and poppy cultivation continue to be a source of revenue with an increase in production from 2013 to 2014.

Education Evolution Overview

Afghanistan's education system is full of changes and controversies with influences coming from the Afghan leaders, the political arena, religious leaders, tribal leaders, and dozens of countries—offering both support and sometimes forced changes. In order to address current topics in mathematics education in Afghanistan, it is necessary to have a basic foundation in the history of education in Afghanistan; i.e. each historical context has and educational consequence that cannot be ignored.

Considering Afghanistan's tribal culture and history, the effects of actions from hundreds of years ago still cause deep resentments today.

In 642 CE, Arab Muslims brought Islam to the region of Afghanistan. By the turn of the 11th century, the last remaining Hindu rulers were defeated and the area of Afghanistan was essentially Islamized.

With Islamic civilization and culture entrenched in the region, many great architectural monuments and mosques were built (Qubain, 1966). Mosques served to spread the word of God and also allowed citizens to learn and serve; indeed, the Islamic civilization produced centers of learning in the major cities such as Kabul, Herat, Kandahar, etc. These establishments were welcomed and people would travel from near and far to study from the educated scholars of the time (Matthews, 1949).

Maktabas (libraries) and madrassas (Islamic Religious Schools) numbered in the thousands and students would learn to read mainly from the Holy Quran. The teachers were mullahs and maulvis and education centered on Islamic traditions and law. In addition to teaching the fundamentals of Islam and the Quran, there were more advanced learning centers located in some of the major cities such as Herat, Balkh, etc. At the advanced centers, students were taught Islamic Law, Theology, Logic, Medicine, Literature, and Grammar (Baiza, 2013). Because these schools offered unique opportunities to get an education, people would often travel from one madrassas to another to get the best education from each resident scholar. This system worked extremely well because it not only served the immediate community but also allowed outsiders to gain an education too. Entire generations of doctors, judges, geographers, mathematicians, poets, historians, etc. were educated as a result of these schools.

Considering that education was a casual, non-mandated process during this time, and also the tribal nature of Afghan society, many students stopped getting an education after obtaining knowledge in the basic tenets of Islam. The reason students would stop their education was the belief that by studying beyond this level, they would not remain good Muslims. Aside from skills such as painting and calligraphy, these schools emphasized morality with students learning the virtues of respect, kindness, generosity, etc. while creative thinking and intellectual curiosity were nonexistent. These notions of Islam forced schools to eliminate the rational sciences from the curriculum in favor of religious education. Very few students studied medicine, geometry, or other natural sciences. Because of this religious hegemony, many schools would produce mullahs and maulvis to work in mosques and madrassas but also to work and serve in the government.

In 1875, Amir Sher Ali shifted education from having solely religious considerations to instead operating with the intention of producing bureaucrats and teachers to manage the affairs of the state. He established two modern schools that were not associated with any maktabas or madrassas: 1) Civil School of the Chieftains, and 2) Military School (Baiza, 2013). These modern schools were not public access but instead admitted the children of the tribal chieftains and also the children of the military elite. The curriculum at the Civil School (a primary school) provided education in reading, writing, mathematics, Islamic education, and English. The curriculum at the Military school is unknown but the teachers came from British India. Curiously enough, a British India general trained the new Afghanistan army. This level of trust and collaboration is testament to the working relationship with the British before the second Anglo-Afghan war. This education system remained up to the 20th century.

With modern changes happening in Europe and other countries in and around the 20th century, educated Afghans worked towards progressive changes in education. The introduction of modern education at this time was not uneventful. The mullahs and maulvis showed a hostile attitude at the reformation of the education system but eventually, a modern system was adopted to create a society that was more globally competitive. Perhaps the biggest achievement of the state at this time was to allow girls to get an education for the first time. This does not mean that girls began getting an education immediately; they did not, however, the idea was planted and eventually girls were allowed to get an education. Supported by religious rhetoric, the battle against girls getting an education continues to this day in the 21st century.

British Rule and Influence in the 19th Century

There were three major conflicts of the British Anglo-Afghan wars (1839–1842, 1878–1880, and 1919) the former was considered the beginning of The Great Game—a power struggle for influence in Asia between the United Kingdom and Russia.

After the second British Anglo-Afghan war ended in 1880, and with Russia ending a war with Turkey two years earlier, the British and Russia aligned their views with keeping Afghanistan as a buffer state (Barthorp, 2002). Abdur Rahman was recognized as the Amir of Kabul in July 1880. As the feud between Britain and Russia eventually diminished further, this lead to the current demarcation of the boundaries of Afghanistan. Unfortunately for the Afghans, these divisions of new states divided the people of similar culture, tradition, and language in pieces with no common lineage or history (Hopkins, 2008). These somewhat arbitrary divisions were part of a larger political plan. The idea being that by dividing particular groups/cultures, the British and Russians would weaken the nation so that it did not pose any threat to interests in India and Central Asia (Hopkirk, 1992). Viewing these divisions through a modern lens, it is perhaps surprising that Amir Abdur Rahman accepted the lines of demarcation and that he did not fully realize the cultural implications that were at stake. The Khyber Pass and the Durand line still have both political and military implications; moreover, the tribal elements and tribal powers that these lines sought to eliminate still exist today and make it more difficult to unify Afghanistan under a single government.

In 1885, under the pretenses of border security and Russian encroachment, the Amir settled 18,000 Pashtun families in Badghis province and in northern Afghanistan. These settlements caused the displacement of tens of thousands of Shias and Hazaras to India (now Pakistan) and Iran (Baiza, 2013). This act caused further resentments among the various tribes that are still strongly felt in Afghanistan today.

It seems safe to say that the mathematics education curriculum in the 19th century was limited to arithmetic, algebra, logic, and geometry with no girls receiving a formal education. Aside from the political disruption of boundary lines being changed and several wars, the British did not intentionally change the curriculum of mathematics and science at this time. Their motivation seemed to be driven by political influence in Central Asia and countering Russian influence at that time. The British did not view education as a tool for war and educating the Afghans was not on their agenda. Ultimately, educational influence did come from the U.K. but it also came from Iran, Pakistan, and more generally Europe.

Early 20th Century Influences to Education

In 1901, Amir Habibullah Khan, the son of Amir Abdur Rahman, began his rule. He was handed a peaceful nation that was already under autocrat rule and had excellent foreign policy relations with Britain. Because the government foundation was already established, Amir Habibullah Khan did need to exercise the strict control over the land as his father did before him; and, unlike his father, he saw the importance of education.

Amir Khan established The Habibian Blessed House of Knowledge (Habibia School) for boys in Kabul in 1903 and is credited with establishing modern education in Afghanistan (Baiza, 2013). It is no surprise that this period also marks the beginning of the divergence from the more traditional madrassa and maktabas education. The new education system was to be framed from the European model of education.

The new system of education would also reduce dependency on tribal leaders for their tribal armies because now, the government would train its own army. By lowering the dependence on the tribal regions, this, in effect also lowered the tribal leader's political clout and placed their religious influence in jeopardy. To appease the resistance, the Amir established 60 primary schools in mosques and hired the local mullah to teach the primary classes.

The Habibia School started as a primary school. Eventually, the school developed into a secondary school which offered three years of primary education, three years of lower-secondary education, and four years of secondary education. After this, six branches of the school were opened in six different districts in Kabul with over 1,500 students enrolled (Baiza, 2013).

With respects to mathematics education, the civil schools offered "mathematics" at the primary and lower-secondary level while the upper-secondary level offered "algebra," "geometry," and "analytical geometry." The mathematics curriculum for the military school consisted of the topics "mathematics" and "geometry" only (Baiza, 2013). These secular topics had a lower status in education compared to the religious topics and in fact, transfer of levels was strictly based upon a student's performance in Arabic and the other religious courses.

Amir Habibulah also established the Education Board in 1904, which served as the precursor to the modern day Ministry of Education or M.O.E. The Education Board was responsible for the governance of the education system including enrollment, curriculum, management, and testing. The international influence on the early education system in Afghanistan becomes stark when you consider that the members of the first Education Board consisted of three Afghan educators, three Turkish educators, and four educators from India. This board established the first Teacher Training College in 1912 with the aim to train primary school teachers.

Because of his loyalty to Britain, a member of the National Secret Association assassinated King Habibulah Khan in 1919.

In 1919, King Amanulah Khan (son of Amir Habibullah Khan) signed the Treaty of Rawalpindi and declared Afghanistan a sovereign state. Amir Amanulah was progressive in his thinking and worked to establish diplomatic relations in the international community (Barthorp, 2002). He also made education a high priority and introduced reforms with the intention of modernizing the country. Among these reforms was to make elementary education compulsory, abolish slavery, and to allow girls to get an education.

Amir Amanulah greatly expanded the education system into a national education system to include primary, secondary, vocational, occupational, and higher education schools (Baiza, 2013). Two new schools were opened in Kabul in 1921 and 1924 that were taught by French and German professors that taught French and German as foreign languages. There was also talk about establishing what would be the first modern university in Kabul but, near the end of his reign, there was considerable conservative opposition and the idea was tabled. By 1929, it is estimated that 51,000 boys were attending 322 primary schools and 3,000 males were attending either secondary, vocational, or occupational schools throughout the country (Baiza, 2013).

Very little documentation exists about girls getting an education at this time. It is known that the first official girls primary school, named Masturat for "purity," was opened in Kabul in 1920. Queen Suraya was responsible for supervision of the school and the women teachers came from the upper class of society. The main subjects were general literacy, Persian, and religious education. A second girls school was opened sometime between 1921 and 1924 in Kabul as well. The estimates are that between the two girl's schools, there were 290 girls receiving a primary education while about 10 were at the lower-secondary level (Baiza, 2013). This is a good time to note that girls have always been allowed education in mosques and in-home tutoring. It is the idea of a state-sponsored education program being open for girls that is modern to this education system, culture, and government.

From 1924 to 1929 the girls' schools were closed and reopened twice due to political opposition. In 1928, Amir Amanulah supported co-education for boys and girls, and girls were being sent to Turkey to pursue a secondary education. At the end of 1928, it is estimated that 800 female students were receiving an education.

The king also was very interested in teacher education programs and issued a law to regulate teacher training in men and women. Books were also translated and published on various subjects including grammar, theology, history, and science (mathematics) to be used by teachers and for the training of teachers (Baiza, 2013).

While there were many reforms during this period, there is no evidence to suggest that changes to the curriculum occurred. The schools still offered mathematics, algebra, geometry, logic, and analytical geometry as the mathematics topics across the curriculum.

Eventually, Amir Amanulah Khan's modern ways became too much for conservatives and even the moderates. By abolishing the traditional burka for women and also opening a co-gender educational school, the Amir alienated the religious leaders. He was forced into abdication from civil war in January 1929 (Barthorp, 2002, Hopkirk, 1992). At that time, all progress in education from the previous 30 years collapsed.

In October 1929, Muhammad Nadir defeated Habibullah Kalakani and became the new king of Afghanistan. His approach favored the Pashtuns and thus he was murdered in November 1933. His son, Muhammad Zahir ruled from 1933–1973 and was deposed by his cousin Muhammad Daoud in a bloodless coup by the Soviet-backed People's Democratic Party of Afghanistan (PDPA) in 1973. Daoud adopted the title of "President" and ruled for five years. In April 1978, President Daoud was killed in a military coup by—again—the Soviet-supported PDPA.

Education during this period (1929–1978) was very political and the main concern was to appease the conservative religious leaders. It was during this period that the state began using education as a political instrument to nationalize Pashtu and eliminate Persian in the country. Ethnic minorities including the Shia and Hazara were banned from higher education institutions and military schools and education for girls ceased to exist. All girls were pulled from Turkey and required to wear veils. In 1933, there were only 27 all-boys primary and secondary schools with about 4,600 students enrolled. In five short years, Afghanistan went from having 54,000 male students to less than 5,000 (Baiza, 2013).

In 1932, Afghanistan founded its first higher-education program by converting a school of medicine into a Faculty of Medicine. Its role was to provide a higher level of medicinal studies than that being currently offered.

In 1945, the MOE founded Kabul University and many of its specialists were filled with foreign academics. This marked the beginning of organized higher education and courses were provided in law, natural sciences (mathematics), literature, and political science (Baiza, 2013).

Afghanistan joined the United Nations in 1946 and because of this, the state was obliged to follow U.N. conventions. In 1949, a midwife school was converted to a 9th grade secondary school for girls with female French teachers. In the early 1950s, the government started a mass education program outside of Kabul. Village schools had a three-year curriculum and were one-teacher classrooms. Most of these schools were in mosques and taught by mullahs. If enrollment reached over 50 students, the school would be promoted to a two-teacher village school. Only after enrollment reached 100+ students was the school to be considered a regular primary school. Similar schools for girls were not established until 1957. By 1970, there were about 1,600 village schools for boys and about 230 village schools for girls (Geneva, 1972).

In 1954, Teachers College of Columbia University (TCCU) was contracted by USAID to provide specialists to develop a program for primary school education. At this time, there were only 13 primary schools for girls and no rural ones; the total number of girls enrolled in grades 1–6 was 5,184. The TCCU team made syllabi for social studies, science, mathematics, and professional content. From 1964–1967, there was a Mathematics Education specialist working in Afghanistan. TCCU left Afghanistan in 1967 (UNESCO, 1972)).

In the 1960s, there were five more school for higher education built: 1) Institute of Industrial Management in Kabul, 2) University of Nangarhar, 3) Higher Teacher Training Colleges in Kabul, Nangarhar, Herat, and Balkh, 4) Academy of Teachers' Education, and 5) Kabul Polytechnic Institute (Baiza, 2013). Sometime between 1960–1961, all institutes for higher education became co-educational and while many challenges still remained, this was a positive step forward for women and education (Tranberg, 1966)

In 1969, TCCU returned to Afghanistan to work with faculty at Kabul University. The focus was to establish programs for teacher education and also a degree program for school leaders and future administrators. In July 1978, the MOE assigned TCCU the additional responsibility of developing a teacher education program for the training of mathematics, science, and social science teachers. TCCU developed printed curricular materials, lesson plans, tests, and also developed plans for future Faculty of Education (AID Project, 1970). During this period, the TCCU team completed the translation and printing of four chapters of materials for the mathematics course: 1) mechanics-impulse and momentum, 2) circular motion, 3) rotation, and 4) harmonic motion. They also completed a first draft of a textbook on Mathematics for Chemistry and Biology students. A report by TCCU makes clear that there were many challenges to overcome during this period; understaffing of mathematics faculty continued to be a "major problem" and at one point, students reacted negatively to a mathematics course and boycotted the class (AID Project, 1970).

TCCU also worked with the Ministry of Education to produce a 1st grade Pashto textbook and teachers' guide that was tested in 1970. A Dari edition of the text followed shortly thereafter. Upon completion of these mathematics textbooks, TCCU wrote 2nd grade, 3rd grade, and 4th grade textbooks (See Appendix). Although it was documented that the 1st grade textbook was pilot tested, it is not known the full extent to which these textbooks were used.

On August 23, 1974, President Daoud changed the length of schooling from six years primary, three years secondary, three years upper secondary (6, 3, 3) to eight years primary, zero years secondary, four years upper secondary (8, 0, 4) (Education in Asia, 1975). He also required graduates from primary school to take an entrance examination (Concours) to qualify for four years of academic schooling (Baiza, 2013). Placement of students who failed the Concours became a problem. If a student failed the Concours, they were promised a one-year vocational school but this did not always happen and instead students were told that they would have to go find a job.

In 1977, the government established the Ministry of Higher Education and began affiliating its institutions with several European and American institutions including collaboration with the University of Wyoming's Engineering Team (Baiza, 2013).

During this period, Persian became the national language. Mathematics education was given a boost with the collaboration of an Engineering School from the United States. This situation remained unchanged until President Daoud's assassination in April of 1978.

Soviets and Mujahidin (1979–1992)

The PDPA took control of Afghanistan after the assassination of President Daoud. This marked a major shift in politics from a Monarch rule to a Marxist-leaning political system (Giustozzi, 1999). The PDPA had strong resistance from lesser-known Islamic parties so in 1979, on the brink of collapse, the PDPA called on the Soviet Union for military intervention. The

Soviets invaded Afghanistan in December 1979. Because the United States declared Soviet occupation illegal, Afghanistan entered into another "Great Game."

Once again, current educational policies were discontinued and education became polarized and used as a political tool. On one hand, the PDPA and the Soviets integrated socialist ideology in textbooks and teacher training; on the other hand, the resistance (mujahidin) developed textbooks with clear, anti-Soviet and anti-PDPA messages (Baiza, 2013). For example, the author has seen mathematics textbooks developed by the resistance that contained word-problems similar to the following:

> "A Soviet helicopter carrying 11 soldiers is shot down and 5 are killed. How many remain *to be killed*?"
>
> or
>
> "You have 5 bullets in your weapon and 9 Soviet soldiers are attacking. How many more bullets are needed *to kill all the soldiers*?" (Kennis, 2006)

This period experienced dual education systems, whereby one was administered by the State and the other by the resistance; both of which were extremely politically charged.

The Democratic Republic of Afghanistan (DRA) did adopt secular policies to education. During this period, girls and woman's access to education was expanded. Pashtu courses were eliminated and the concours were abolished (Baiza, 2013). The DRA also encouraged free education for all regardless of gender, religion, or language. This type of thinking went along with Marxist ideals of freeing the oppressed including women. To date, this period (1978–1992) shows the highest level of female participation in education and public services.

Although women and minorities seemed to show gains in the political and educational arenas, the enrollment for all-boys schools fell from 44% enrollment in primary school (1975) to 27% (1985). The number of schools (primary and secondary) decreased from 3,352 (1978) to 586 (1990). Girls enrollment went up in the urban areas and stayed the same in the rural areas. The reasons for these statistical changes are clear. While the DRA has high ideals to educate all people, particularly the oppressed, implementing these ideals in the rural areas during a civil war proved difficult if not impossible (Baiza, 2013).

Throughout the 1980s education was a political tool to achieve the party's political goals. Social Science and History textbooks were rewritten and filled with topics about revolutionary movement, labor movements, British occupation, etc. In 1986, MOE approve a new subject called "Patriotic Military Education" for boys in years 10 and 11 in upper-secondary courses (Baiza, 2013). The idea behind this course was clearly the enculturation of boys into the military.

Political arrests and executions of teachers and educators were so rampant in the late 1970s and early 1980s that Kabul University had to close its Faculty of Engineering in 1985 because of lack of lecturers (Baiza, 2013). Clearly this dealt a significant blow to the advancement of mathematics education at that time.

The war-education that was delivered by the Mujahidin between the years 1984–1987 via textbooks and through biased teaching was limited to non-Shia schools. When the Taliban emerged to power in 1994, there were thousands of teens that had been indoctrinated and radicalized by this system.

Seemingly, every regime change scraps the old education system for a new one. This lack of continuity in the education sector has produced a drag on the system that is observable today.

Post-Soviet Taliban Era (1992–2001)

When the Soviet's withdrew from Afghanistan, the resistance parties' Islamic State of Afghanistan (ISA), as lead by Ahmad Shah Masoud, went immediately into a civil war that lasted until the Taliban drove them from power in 1996. During the four years before the Taliban, the education system became further fragmented due to the civil war. While the ISA was not opposed to a systematic education system, they wanted to Islamize the system. This meant to eliminate the current policies and programs. New textbooks and curriculum would be written with Islamic principles and Islamic thoughts (Baiza, 2013). Girl's education would be in line with the needs of an Islamic society. In Kabul, girl's education was allowed but co-education was completely eliminated.

From 1992–1996, USAID assisted the government with a new curriculum. As the education system continued to crumble and rural education offices were controlled by different leaders, the government lost control of the education system. As such, the USAID project and curriculum work was never implemented. During this time, the education system lost all supplies and support and—aside from textbook writing—no new developments were considered (UNESCO Islamabad, 1996).

There were several warring factions involved in the civil war and each had their own quasi-government with separate ideas for education. Even the city of Kabul was divided among several resistance groups—the Uzbek, Tajiks, Pashtun, and Hazara. The Taliban were initially seen as the group to bring stability and peace, however, it became clear that they were just another fighting force struggling for power.

The leaders of the Taliban forces were largely resistance fighters from the Soviet-era. Their soldiers or underlings were, by and large, students of madrassas who studied in the Pashtun areas of Pakistan. Their curriculum would consist of religious extremism and hatred for the Soviet occupiers. Their textbooks were that of the resistance movement which, as previous described, contained clear, extreme political views and violence. Almost needless to say, the Taliban strongly aligned itself with the religiously conservative madrassa education system.

The infrastructure to the education system was decimated before and during the time that the Taliban took control in 1996. While Kabul managed to avoid major damages during the Soviet war, the damage to Kabul's infrastructure during the civil war and the Taliban takeover was unprecedented. Virtually all schools were destroyed or damaged in some way. It is estimated that 60% of the educational infrastructure was destroyed in Kabul and 70% of all educational buildings were destroyed in the provincial areas; these statistics include higher education school, and also vocational, technical, and teacher training institutions (UNESCO, 2000).

Under the Taliban, government services collapsed and all girl's schools were closed and women's restrictions to education were, once again, imposed. Any boys' schools that were still standing were transformed into madrassas. Children's education, traditionally handled in the mosques, was controlled by the Ministry of Religious Affairs. By April 2000, all female teachers were removed from the education payroll and girl-friendly NGO's were expelled (Baiza, 2013; UNESCO, 2000). By 2001, the entire modern education system of Afghanistan—beginning in 1901—had collapsed to nonexistence. The country's education system and culture was set back 100 years.

Reconstruction and Democracy

With the fall of the Taliban in November 2001, the Bonn Agreement was quickly composed with the support of the UN and many global leaders on December 5, 2001. Very broadly speaking, the Bonn agreement set the foundation to allow Afghanistan to reestablish itself politically as a state and declare an interim President (President Karzai). By mid-2002, organizations such as the Asian Development Bank (ADB), the Afghanistan Assistance Coordination Authority (AACA) and others developed Comprehensive Needs Assessment (CNA) and also educational policy to set the stage for education priorities and strategies (Sarvi, 2003). In 2004, the Ministry of Higher Education (MOHE) and UNESCO's International Institute for Educational Planning (UNESCO-IIEP) developed the first Higher Education Strategic Action Plan (NHESP) (Baiza, 2013; Sarvi, 2003).

Curriculum reform was necessary and UNESCO spearheaded a textbook development and curriculum reform project in 2002 (Baiza, 2013; Sarvi, 2003). Teachers College at Columbia University was among the many international, educational support teams that sent curriculum specialists and textbook writers to assist. The Teachers College/Columbia University Afghanistan Project (TCCU-AP) sent specialists in many areas of education including English, Literacy, Curriculum Development, and Language and Cognition. In 2004, the author was the Mathematics Education Specialist for this team and quite possibly the only mathematics educator working under UNESCO in Afghanistan at this time.

The role of the mathematics education specialist under TCCU-AP was to write a 2nd and 3rd grade mathematics syllabi (textbook) collaboratively with the Afghan Mathematics Team in the MOE. Although this work was already completed in the 1970s by TCCU, no one in the MOE knew where the work was stored or if it was even available. A secondary role is to teach a couple of workshops to pre-service teachers at Sayyid Jamaluddin Teacher Training School. The working office would be located in the MOE in the Compilation and Translation Department.

The Afghan mathematics team in the MOE consisted of Mr. Abdulkabir, Mr. Sher Ahmad Gardiwal, Mr. Nazamuddin, Mr. Shah, and Mr. Boba Saheb Tufan. Mr. Gardiwal also worked as one of the team's translators and as a Biology Professor at Kabul University. Morale was low because these men made the equivalent of $35/month working for the MOE and they can make much more selling vegetables in the market. To complicate matters further, initially TCCU-AP had language specialists working with the math team so frustrations were high. The mathematics team was grateful to have a math colleague to work with and the author would meet with these men daily whenever possible.

The mathematic syllabi requested by UNESCO and TCCU-AP were not syllabi in the traditional sense; instead, these were complete textbooks with problem sets, instructions for the teacher, and hands-on activities that required nothing more than sticks and rocks to implement. By having 140 lessons, the material would contain more overlap of previous lessons than the previously used books; for example, instead of one lesson about the sum of single digit numbers, two lesson were written where the sum of the digits is less than ten and then also the sum of the digits is greater than ten.

As with most curriculum projects during that time, the UNESCO deadlines did not leave enough time to fully interact and completely collaborate with the Afghans—a necessary component of international development. As pages were written, they were translated and then given to the Afghan mathematics team for input and support. The syllabi were completed and turned over to the MOE and UNESCO in December 2004 when the author left Afghanistan. The rest of the TCCU-AP team returned to New York City in 2005. In 2010, the MOE completed the entire curriculum development and also teacher guides and textbooks for grades 1–9 (MOE, 2004). It is not known if the grades two and three syllabi from 2004 were ever used in any capacity.

Teacher training during the reconstruction period was highly necessary not only because of the large increase in student enrollment but also because the education level of the current teacher base was quite low (60% had only 12th grade credentials) (MOE, 2003; Georgescu, 2007; Zoepf, 2006; Morrison, 2002; Husting 2008). The author had the opportunity to teach two workshops at the Sayyid Jamaluddin Teacher Training School, one of which was particular to mathematics education. The topic was "children's arithmetic mistakes" and the students (about 60% women) were very eager to learn; participation level was high. It was observed that while the pre-service students certainly knew the arithmetic, they had almost no problem-solving or critical-thinking skills. One can only imagine that after 30 years of non-stop wars, those skills did not receive the proper nurturing or education to fully develop.

2015 and Beyond

According the Afghanistan Ministry of Education's website (www.moe.gov.af), when new government was formed in 2002, the education system was in distress:

- There were less than one-million students, 20,000 students and almost no females.
- There were 3,400 schools and most were destroyed/unusable.
- There was no standard curriculum and no standardized textbooks.
- There were four teacher training institutions with only 400 students.
- There were only 1500 boys enrolled in Technical and Vocational Schools.
- There were 220 unregulated madrassas without any curriculum.

In 2015, the statistics are impressive and a testament to the amount of work that the Afghans and donors have done:

- There are 7 million children enrolled in schools with about 2.5 million girls.
- Over 4,500 schools have been built with active community involvement.

- There are 170,000 active teachers with 30% being female.
- There are now 42 Teacher Training Centers and 73 district Teacher Training Centers with a total of 42,000 students enrolled (38% female).
- There are 60 Technical and Vocational Schools with 20,000 students (3,000 girls).
- There are 42 Centers of Educational Excellence and about 485 registered madrassas.
- There are over 8,500 school Shuras (councils).

Ghulam Faroug Wardak, the current Minister of Education, outlines the MOE's plans in the National Education Strategic Plan (2015–2020) (NESP III) in accordance with Millennium Development Goals (MDGs) and Education for All (EFA) goals. Among these goals is to ensure that all children, everywhere, boys and girls, will be able to complete a full course of primary schooling by 2020. This comprehensive plan is divided in two scenarios: low-case and high-case. The low-case is based upon realistic estimates and budget constraints, the high-case scenario is designed to meet the MDGs and EFA objectives. With respects to mathematics education and science education, the MOE states as a strategy that,

> The MOE Science Center will improve the quality of science and mathematics education in schools by training science teachers on the utilization of laboratories for implementing scientific experiments, distributing teacher guides for scientific experiments, equipping schools with laboratories, and distributing mathematics kits and other learning materials.

This is the only mention of the word "mathematics" in NESP III and it clearly leans towards science as a discrete subject. As such, teaching mathematics in Afghanistan will remain a "rote" topic to be memorized and repeated. Until the MOE recognizes mathematics education as a vehicle for critical thinking and problem solving skills, Afghanistan will impede its own progress as a creative, independent nation.

With a nationalistic yearning for education and global support, Afghanistan is on the road to educational stability. Political stability in the region always poises threats to the system, however, students now receive

12 years of mathematics classroom education and can pursue higher education degrees in mathematics, engineering, and medicine in several four-year institutions; a truly amazing feat for Afghanistan and the global community.

References

Baiza, Yahia. Education in Afghanistan: Developments, Influences and Legacies since 1901. London and New York: Routledge, 2013. Print.

Barthorp, Michael. Afghan Wars and the North-West Frontier: 1839–1947. London: Cassell &, 2002. Print.

"Basic Education Strategies Afghanistan." UNESCO Islamabad 1, 1, August 1996: Print.

"CIA World Factbook." Afghanistan. Web. 17 Sept. 2015.

"Country Education Profiles: Afghanistan." International Bureau of Education: Geneva, June 1972: Print.

"Curriculum Framework." Afghanistan Ministry of Education Compilation and Translation Department 2003. Print.

"Development of Education in Afghanistan during the Period 1960–1970." Bulletin of the UNESCO Regional Office for Education in Asia 6.2 (March 1972): 1–10. Print.

"Educational Reforms in the Republic of Asia." Education in Asia 8.13–5 (Sept 1975): Print.

Georgescu, Dakmara. "Primary and Secondary Curriculum Development in Afghanistan." Prospects 37 (2007): 427–48. Print.

Giustozzi, Antonio. War, Politics and Society in Afghanistan, 1978–1992. Washington, DC: Georgetown UP, 1999. Print.

Hopkins, B. D. The Making of Modern Afghanistan. New York: Palgrave Macmillan, 2008. Print.

Hopkirk, Peter. The Great Game: The Struggle for Empire in Central Asia. New York: Kodansha International, 1992. Print.

Husting, Sheila, Joann Intili, and Edward Kissam. "Teacher Training in Afghanistan: Intersections of Need and Reality." Convergence 41.2–3 (2008): 27–40. Print.

Jr, Carl C. Tranberg. "Mathematics Teaching in Afghanistan, 1964." *The Mathematics Teacher* 59.2 (1966): 143–46. Print.

Kennis, James. "Afghanistan: Mathematics Education Reconstruction Effort." MAA Focus. 1st ed. Vol. 26: Mathematical Association of America. Print. January 2006.

Matthews, Roderic D., and Matta Akrawi. Education in Arab Countries of the Near East. Washington, D. C.: American Council on Education, 1949. Print.

Morrison, Bronwen. Proc. of Food and Education in the Reconstruction of Afghanistan. Final Proceedings Report: AIA-NGO-WFP-CAII Workshop, 17–20 February 2002 Islamabad, Pakistan. Creative Associates International, Inc., Washington D.C., Web.

Qubain, Fahim Issa. Education and Science in the Arab World. Baltimore: Johns Hopkins, 1966. Print.

Sarvi, Jouko. A New Start for Afghanistan's Education Sector. Manila, Philippines: Asian Development Bank, South Asia Dept., 2003. Print.

Strategic Action Plan for the Development of Higher Education in Afghanistan. Place of Publication Not Identified: Ministry of Higher Education, Afghanistan, 2004. Print.

"Teacher Education in Afghanistan: Thirty-second Six-month Report / Teachers College, Columbia University, A.I.D Project." A.I.D.Project 306-11-690-091 (1970): Web.

"United Nations Educational, Scientific, and Cultural Organization." World Education Report 2000: The Right to Education (Towards Education for All throughout Life) (New York: UNESCO Publishing, 2000): Print.

Zoepf, Katherine. "Progress and Pain in Afghanistan." Chronicle of Higher Education 52.21 (Jan 2006): 44–48. Print.

About the Author

James Kennis received both the master's degree and doctorate in mathematics education at Teachers College. He served in Afghanistan as member of the Teachers College Advisory Team with responsibility for revising school mathematics textbooks after Afghanistan's return to an elected government. Professor Kennis has further worked as a teacher-trainer in mathematics and science in Ilocos Sur, Philippines and also taught graduate classes in mathematics education at Khon Khen University in Thailand. Dr. Kennis is a member of the mathematics faculty of Hostos College of the City University of New York. His research interests include mathematics remediation, probability theory, and games of chance.

https://doi.org/10.1142/9789813146785_0002

Chapter II

ALGERIA: Algerian Mathematical Practices: Past and Present

Ahmed Djebbar
Professor Emeritus at the University of Lille, France

Introduction

Mathematical practices in Algeria classified by their different purposes (theoretical or applied) and their different forms of expression (oral, written, or artistic) have had a long history that can be divided into four major periods. The first preceded the conquest of North Africa by Muslim armies (642–711). After a period of exclusively indigenous governance, the area became partially occupied and administered by authorities from other Mediterranean cultures: first the Phoenicians (late 9th Century–146 BC) then the Romans (146 BC–429), the Vandals (429–533) and the Byzantines (533–698). In the second period, Algeria became a sub-region of a vast empire that would be governed in the name of Islam (702–1830). This governance was not always "centralized," in so far as a number of regional states, independent of the Caliphate's authority, were formed in the area that corresponds today's Algeria. The third much shorter period was the occupation of the territory by France and the installation of a colony by migration (1830–1962). The fourth and final period is independence, which was won in 1962.[1]

Before the conquests described above, the language of the inhabitants was Berber or "Tamazight." As far as we know, this language did not, during this long period, produce a body of knowledge or scholarly mathematics teaching. Although there is no recorded written work in mathematics education at this time, that does not mean there were no mathematical practices. In fact, research specialists in ethno-mathematics studying the know-how of African societies have shown that concepts and arithmetic and geometric tools originated and were mastered in the context of commercial or artistic activities. Archaeological discoveries and the survival of some current practices (geometric decorations, measurements, numbering and calculations) reveal the same occurrence in Berber societies of the pre-Islamic period.[2]

In this sphere, the situation did not change during the long Phoenician and Roman presence. Other than St. Augustine's reflections (d. 430) on arithmetic, music and architecture, historians report no local scientific production in Latin.[3] St. Augustine was bishop of Hippo which is now Annaba in Algeria. This situation did not change during the occupation by the Vandals. As for the Byzantines, even if they controlled only a small part of what is currently Algeria,[4] one would think that their language, which was Greek, would have been a vehicle for science, or at least for part of the scientific projection of the Hellenistic period (323–30 BC). This hypothesis is based on the status of the Greek language among the elite of some Berber kingdoms particularly those ruled by Massinissa (204–148 BC) and Juba II (25 BC–23 AD).[5] However, we know today that at the time ancient Greek sciences, and particularly in mathematics, were out of favor with Byzantine authorities. In spite of this, it is quite possible that utilitarian mathematics may have been practiced in Carthage. There is evidence of dissemination of a non-positional numeration system using 27 symbols (9 for units, 9 for tens and 9 for hundreds) which was known in the Muslim West under different names (Fez numbers, register numbers, Byzantine calculation). We know today that these 27 numbers are a variation of the Greek alphabet that astronomers used in their calculations and in their development of astronomical tables. We do not know how this numeration was circulated, but its presence and use in Andalus and the Maghreb is proven by the many mathematical or legal documents that have survived.[6]

With the arrival of Islam in the Maghreb, a new civilization developed. The mastery of papermaking and the growth of its production, first in

Baghdad, the capital of the empire, and then in other regional cities, encouraged the circulation of the first scientific publications (translation of Greek and Indian works into Arabic original works). These publications circulated in parts of the Maghreb and gave birth there to a local scientific tradition and in particular, the production of the first mathematical works. Kairouan, in the city southeast of the Maghreb, was the pioneer in this field. Dating from the ninth century, manuals on "Indian computation" and the science of inheritances were published. From the tenth century, other cities in the region followed suit.[7] The content of this production seems to have included both theoretical and applied aspects of mathematics. These activities required the existence of both a basic education and more advanced and targeted education. Their content did not vary much from one scientific center to another. It was defined loosely and directed by teachers who extended their teachings through the publication of introductory manuals or books to meet specific needs.

Mathematics in Central Maghreb

In the context of the civilization of Islam, Algeria represented the historical continuation of a geopolitical entity that was called from the 14th century "central Maghreb". This name came as a consequence of the fragmentation of the Almohad empire (1147–1269), whose decline began after the famous battle of Las Navas de Tolosa in 1212, with the victory of the Castilian army and its allies over the army of the Caliph Muhammad al-Nasir (1199–1213).[8] Before this period, different regions of central Maghreb had known various destinies on the political level: integration of some into large States such as the Almoravid empire (1040–1147) and the Almohad empire or allegiance to regional powers, such as the Idrisid (789–985) and Aghlabid (800–909) kingdoms or later, the Caliphate of Córdoba (929–1031). Some managed to acquire complete autonomy. This was the case, in particular, with the two City-States of Tahert (767–909) and the Qal'a of the Banu Hammad (1014–1152). These two City-States probably facilitated the slow emergence of the Ziyanide State that had Tlemcen as its capital and lasted longer than its predecessors, about three centuries (1240–1554).[9]

On the economic front, Central Maghreb had taken advantage of its geographical position, with its large land trade routes, its coastline and its Saharan borders to acquire an important place in the international commercial network controlled from the late eighth century, by various authorities of the Muslim empire. The increasing wealth of this area enabled the emergence and development of a number of cities that became vibrant intellectual centers. The most important were Tahert, the Qal'a of Banu Hammad, Ashir, Bejaïa, Tlemcen and Constantine. Other cities, smaller in scientific output, were known by the names of the men of science who came from them. This is the case of Biskra and Tobna, at the entrance to the Sahara, Buna (now Annaba) and Wahran (Oran) along the Mediterranean. In varying degrees and at different periods of their history, each of these cities participated in mathematical activities that developed first in the East, from the late eighth century, before continuing in other regions of the Muslim empire, particularly in the Maghreb.[10]

The Period of City-States

Beginning in the eighth century, Tahert and Tlemcen distinguished themselves by becoming centers of power, completely autonomous from the Caliphate of Baghdad. Concerning Tahert, existing sources describe its leaders as people who all had solid training in various religious and non-religious fields such as mathematics. This is not surprising considering the close relations that the elite of this city had maintained with the elite of the center of the Muslim empire, in Baghdad and Basra.[11] This dynasty's historians mention the solid education of the first Imam Ibn Rustum (776–787) first in Kairouan (the first scientific center of the Maghreb) and then in the East.[12] With the Imam 'Abd al-Wahhab (784–823), intellectual activities benefited from the economic development of Tahert and the increasing wealth of its population, resulting particularly from the control of part of the trade with the African Sahel.[13] They also benefited from the privileged ties established with the strong Ibadi communities of the East that shared the same religious orientations of Ibadism Tahert. With the coming to power of the Imam Aflah (871–894), the dynamic scientific movement amplifies. Copies of the first mathematical and astronomical works produced in the East were available in the city, probably in the famous library of the Imams, called al-Ma'suma.[14] It is also known that the Imam himself benefited from good training in computation and astronomy. Referring to this training, his biographer wrote "that he had reached in Ghubar computation and astronomy a significant level".[15]

With regard to the city of Tlemcen, we know nothing about its first contributions to science and, in particular, the teaching and mathematical production from 786 the date of its foundation by the Banu Yafran, up to 931 the year of its control by the Fatimids. But taking into account the development of its religious practices, particularly those that required a minimum of knowledge in computation and astronomy, it is likely that the city had experienced this type of teaching. This is what immediately comes to mind when one reads what is reported in a chapter of the book by Yahya Ibn Khaldun (d. 1378) about the scholars of Tlemcen. Regarding his city's cultural past, the author refers to "What it produced as expert students, as wise men and as talented craftsmen in each discipline," adding "And if we had tried to mention them all, catalogs would not have been sufficient to contain those among them whom we have information about, the most numerous of whom being from bygone eras."[16]

With regard to the 10th–11th centuries, information provided by bio-bibliographic sources tend to show that most scholars or men of letters of the central Maghreb of this era had a career mainly in Andalus, in the East, or in cities of Ifriqiya. This was the case of al-Wahrani (d. 1037), a man of science best known for his "powerful knowledge in computation and medicine" and who lived mostly in Andalus.[17] It is also the case of Ibn Abi l-Rijal (d. 1035) who made his career in Kairouan and is described by his biographers as simultaneously a man of letters, a scientist and a clerk of the State. He was in fact the Secretary of the Zirid prince al-Mu'izz (1016–1062) and, as such, he developed patronage for literature and the sciences. His only known work associated with astronomy is his famous treatise of astrology, *al-Bari' fi ahkam al-nujum* (The Outstanding Book on the Judgement of the Stars) that became a great success in Europe after its translation into Latin.[18]

Available sources of the 12th century show that a large number of men of letters, grammarians, linguists and jurists came from central Maghreb. Some of them were qualified specialists or experts in mathematics, in particular in the science of computation that was part of the training of judges and their collaborators.

The most dynamic city of that time for intellectual activities in the broad sense was undoubtedly Bejaïa. It held a strategic position as a Mediterranean port and was a compulsory stop on the pilgrimage route to Mecca. This made Bejaïa the meeting place for many students, scientists and intellectuals in various specialties.[19]

Among the men of science of that era born in the region and known for mathematics was Ibn Ma'sum al-Qal'i (d. 1156), a specialist in computation.[20] He was educated in Qal'a of Banu Hammad, the capital of the Zirid Kingdom (972–1148) before settling permanently in one of the cities in Iraq. Sparse information suggests that this brief capital was a scientific center. The testimony of the historian Ibn Khaldun (d. 1406) leads in that direction. He refers to Hammad, the founder of the dynasty, saying: "He planned and located the city of al-Qal'a on the Kutma mountains in the year 398 . . . He had buildings and walls built there . . . Men of science and master craftsman came there from remote cities and regions to enrich the centers of science, trades and the arts."[21]

The second intellectual, known for practicing mathematics, was Yusuf al-Warjalaṇi (d. 1174), an Ibadi theologian. Among the books he published, was one entitled *al-Dalil wa l-burhan* (*The Argument and the Proof*) that features, in its second part, definitions and concepts essential for training in arithmetic and geometry.[22] The third scientist, Ibrahim al-Tilimsani (d. 690/1291), is a lawyer described as an expert in "the science of numbers" and "calculation of inheritances." He acquired notoriety at the age of twenty, after the publication of a treatise in verse on the distributions of inheritances.[23] This poem was of interest to many commentators, including three mathematicians from Central Maghreb: Ibn Qunfudh (d. 1407), from Constantine, al-Habbak (d. 1463) and Ibn Zaghu (d. 1441) from Tlemcen.

Mathematics of Bejaïa

The members of the second category of scientists of this period lived in central Maghreb, most often in Bejaïa. Abd al-Haqq Ibn Rabi (d. 1276), was a jurist from al-Andalus who was well versed in the science of computation and of inheritances.[24] Abu 'Abdallah al-Qal'i (d. about 1266), was the author of a treatise on inheritances, entitled *Nihyat al-qurb* (The ultimate proximity). We know him from the testimony of al-Ghubrini, a biographer who said of him: "He had a theoretical and practical knowledge of law and estate sharing and had a greater knowledge of the computation than <that of> the Ancients . . . There was nobody in Bejaïa, in his time, who, wanting to study this science, would not have studied with him; and people came <from other countries> to study this science with him."[25]

The third, and most important, mathematician of this period was Ab l-Qasim al-Qurashi (d. 1184). He was from Seville where he taught for some time before settling permanently in Bejaïa. He was known for two of his publications that deal with algebra and the science of inheritances.[26] His treatise on algebra was a commentary on the *Kitab al-Kamil fi l-jabr* (*The Complete Book of Algebra*) by the Egyptian Abu Kamil (d. 930). Only a few excerpts from al-Qurashi book have survived.[27] His book on estate sharing has not yet been found but we are well informed about its content and its original suggestions concerning the introduction of a new technique for the calculation of the shares of each person entitled person.[28]

Two other scientists from Bejaïa had a solid mathematical profile. The first is Nasir al-Din al-Mashaddali (d. 1331), who acquired his first training in his hometown before going to the Near East where he studied for twenty years, with great teachers in the region. Back in Bejaïa, he devoted himself to the teaching of certain subjects, especially law, logic and perhaps mathematics.[29] His best known student was Mansur al-Zwawi (d. 1368), described by a contemporary as a scientist with "a good contribution in many rational sciences and transmission . . . as well as achievements in computation, geometry and <the science> of instruments." After training in Bejaïa, he studied in Tlemcen, then moved to Grenada to teach at the an-Nasiriya Madrasa that had just been built.[30] The third practitioner of mathematics was 'Isa al-Mangallati (d. 1342) who was born in the same city as the previous two. He specialized in the science of inheritances and is presumed to have published on the geometry of measurement.[31]

The Mathematics of Central Maghreb in the 14th and 15th Centuries

During these two centuries, the mathematical activities in the scientific centers of Central Maghreb experienced a certain dynamism, which benefited from the decline of the same activities in Andalus, as a result of the continuous reduction of Muslim space in the Iberian Peninsula. Indeed, the control of many Andalusian cities by the Castilian army caused a steady emigration of elites to the cities of the Maghreb. From the 13th century onward, there was throughout the Maghreb a significant increase in the number of practitioners of mathematics (teachers and textbook authors) in applied fields like the geometry of measurement and the calculation of inheritances, as well as in more "theoretical" fields. It seems that this was the result of the coexistence of two mathematical traditions: one that originated in al-Andalus and the other that originated in the first scientific centers of the Maghreb.

Among the initiators of the first tradition, there was Ibn 'Aqnin (d. 1225) in Sebta, Ibn Muncim (d. 1228) in Marrakech[32] and al-Qurashi in Bejaïa. They were joined by scientists who were born and trained in the Maghreb, such as Ibn Ishaq al-Tunusi (about 615/1218)[33] who worked in Tunis and in Marrakech, and al-Qadi al-Sharif (d. 1283–84) who was a pupil of Ibn Muncim[34] and Ibn Haydur (d. 1413), a scientist from Fez. Among the mathematicians of central Maghreb who were linked to this tradition, are al-Abili and al-'Uqbani in Tlemcen and, probably, students taught by al-Qurashi in Bejaïa.

The second tradition in Maghreb probably originated in Kairouan. In terms of content, it is essentially computational and characterized by the development of fractions and the dissemination of arithmetic and algebraic symbolism. But what most characterizes this tradition is the new pedagogical orientation that took place and that was progressively imposed in education and in the editing of textbooks: the suppression of theoretical aspects and in particular, the gradual reduction of demonstration and a greater use of memory. This encouraged the proliferation of short manuals and the memorization of mathematical poems. This trend was exacerbated by the development of the phenomenon called "the ijazah" (diploma). Given by each teacher, these diplomas certify neither a global level nor the mastery of the content of a subject. They certified only the reading and understanding of the content of one or several works by the professor who gave the diploma.

The figurehead of the Maghreb's mathematical tradition was undoubtedly Ibn al-Banna (d. 1321), one of the last innovative mathematicians of the Muslim West. In mathematics, it is mainly his works on computation and algebra that the teachers read and sometimes commented upon. These include the *Talkhîs a'mal al-hisab* (*Summary of the Operations of Computation*), the *Arba' maqalat* (The four Epistles), the *Raf' al-hijab 'an wujuh a'mal al-hisab* (*Lifting the Veil on the Operations of Calculation*) and the *Kitab al-usul wa l-muqaddimat fi l-jabr* (The Book of Foundations and Preliminaries of Algebra).[35]

Some students in central Maghreb had the privilege of following the teachings of Ibn al-Banna or of his students, thus enabling his method and its pedagogy to be disseminated widely. The oldest of his students that we know of is Ahmad Ibn Hammad, a descendant of the Zirid dynasty. He attended classes taught by Ibn al-Banna from 1302 to 1308.[36] But the best known is al-Abili (d. 1356). He was a great teacher of mathematics, philosophy and logic. Originally from Tlemcen where he trained, he belongs to the Andalusian tradition. But as a student of Ibn al-Banna in mathematics, he was also a representative of the Maghrebian mathematical tradition. Among his students from central Maghreb, who received mathematical training from him and who reproduced or continued his teaching in this field was firstly Ibn al-Najjar (d. 1349), an excellent mathematician who would specialize later in astronomy and astrology.[37] The second student was also from Tlemcen, Sa'id al-'Uqbani (d. 1408) who taught logic, the science of inheritances, and mathematics.[38] In this discipline, he is one of the last authors of the Maghreb, with Ibn Haydur (d. 1413), to have taken the trouble to justify the algorithms or the geometric properties which he described using rigorous proofs. He made various references in his commentary to demonstrations based on Euclid's *Elements* (3rd century BC). In this sense, we can connect him to the Andalusian tradition, even if it is through the writing of his commentary on the *Talkhîs*, a link in the Maghrebian tradition that developed from the school of Ibn al-Banna. The third student of al-Abili is none other than the famous ‘Abd al-Rahman Ibn Khaldun. He is best known as a historian but, in his youth, he had the opportunity to publish a paper on mathematics[39] and he taught this discipline for some time.[40]

Among mathematicians whose writings continue the work of Ibn al-Banna, there is Ibn Qunfudh, a scholar from Constantine who did not know the scientist from Marrakech but, in Fez, took the courses of one of his students al-Laja'i (14th century). Like many of his colleagues of this period, he was an eminent jurist, a mathematician and an astronomer.[41] Two other scientists in the same city specialized in astronomy: ‘Ali al-Qasantini[42] and ‘Azzuz al-Qasantini.[43] Their contributions continue the astronomical tradition of al-Andalus and the outer Maghreb, where mathematical concepts remained very important.[44]

The Teaching of Mathematics in Tlemcen in the 14th and 15th Centuries: Institutions, Programs and Pedagogies

We know almost nothing about the teaching of mathematics in the first years of a child's education in central Maghreb during the 14th and 15th centuries. Regarding the second phase of education, which begins at adolescence, its duration ranged between five and fifteen years, depending on the location and, above all, on the pedagogy used. At least this is what Ibn Khaldun said. He counterbalanced the Andalusian pedagogy (which was practiced by some teachers in Tlemcen and Tunis) and the pedagogy that prevailed in outer Maghreb.[45] Concerning mathematics, we have some information reported by biographers or teachers who lived in Tlemcen for some time.

Teaching locations in this city, in the 14th and 15th centuries, included mosques and madrasas. There were about a dozen of the former.[46] Originally, institutions of this type were essentially places of prayer, but between the prayers, they could serve as teaching centers. Some of them had libraries. For the period that interests us here, teachings there given were not always limited to the sciences of the Koran, the Hadith and the Foundations of the Law. There were sometimes courses in *Kalam* [speculative theology], logic, and even philosophy.[47]

On the teaching of mathematics in the mosques, available sources shed no light on this. But that does not mean that there wasn't any. Indeed, it seems inconceivable that they would not offer mathematics courses in institutions that taught logic and philosophy. Furthermore, it should be noted that teaching in Tlemcen was programmed by teachers according to their individual specialties, and not imposed by a state dependent trusteeship or determined by the teaching location. Finally, let us not forget a characteristic of that time: teachers were versatile, with a double or a triple formative education in subjects with completely different contents. So, in this period, scientists most often held double specialties enabling them to teach "rational" sciences, such as mathematics and astronomy, and "traditional" sciences, such as law, theology, and rhetoric. One last argument that supports the claim that there was mathematics teaching in some mosques is the obligatory presence of the science of inheritances in the curriculum for training the elite of the time. This subject is a particularly important component of Muslim law that was always taught in mosques. And, this teaching required corresponding mathematical training involving important components of the science of computation, in particular fractions, their theoretical aspects (integer arithmetic) and practical aspects (conversion, proportional distributions, search for common denominators). It even happened that algebra was taught to students to enable them to understand some of the inheritance problems requiring manipulation of equations.[48]

The second category of institutions of higher education in Tlemcen, i.e. the madrasas, are more important for the disciplines that concern us here. The oldest is the madrasa of the Ibn al-Imam which was built specially for them by the fourth Ziyanid king, Abu Hammu I (1308–1318).[49] This madrasa continued to operate after the death of the two teachers. Two mathematicians studied there—al-Abili and, after him, Sa'id al-'Uqbani. A second madrasa was built by King Abu Tashfin (1318–1337) and it bears his name, *al-Tashfniya*. Students could live there, which the future mathematician al-Masmudi (d. 1401) did.[50] The third madrasa, known as al-Ya'qubiya, was built by King Abu Hammu II (1359–1388) for his son-in-law, the teacher Abu 'Abdallah al-Sharif. It was inaugurated in 1363[51] and remained active until at least 1441. In that year, the mathematician al-Qalasadi (d. 1486) came from al-Andalus and began to attend courses there.[52] Education in this institution gave great importance to the rational sciences, i.e. philosophy, speculative theology, logic and mathematics. Eighty years later, the rational sciences were still taught there, as al-Qalasadi. has confirmed. In his autobiography, he says that the curriculum planned by his teacher Ibn Zaghu (d. 1441) reserved "winter for the exegesis of the Koran, the Hadith, Law and the foundations <of Law>; and, summer for Arabic, rhetoric, calculation, inheritances, and geometry."[53]

Other less prestigious madrasas also operated in Tlemcen. For the most part, they were subject to the same statute, "waqf" (bequests), that guaranteed them sustainable funding from bequests made by kings, princes, or private individuals.[54] Four of them are described from available sources: the madrasas of al-Ubbad, Manshar al-jald, Zawiya Sidi al-Halwi and Sidi al-Hasan.

The Content of Mathematics Taught After the 13th Century

Regarding the overall trends in mathematical activities, our first observation is that the teaching content and the mathematical books gradually get smaller with fewer topics included. We see this also in the profiles of the teachers and the authors of books who are specialists in calculation and its application to the science of inheritances. The teaching of algebra and Euclidian geometry is scarcely found. In the textbooks, some components have disappeared, such as those that discussed number theory problems or relatively sophisticated approximation methods.

In geometry, works of the great Arab tradition of the 9th–12th centuries are no longer mentioned. They have been replaced by basic textbooks presenting the properties of traditional plane and solid figures and solving problems for determining certain elements of those figures by knowing the others. The writings of Ibn Luyun (d. 1349), Ibn al-Raqqam (d. 1315) and Ibn al-Banna represent the practical geometry of that time.[55]

But in spite of everything, Euclid's *Elements* are still the reference, even if we do not know which chapters of the treatise were taught. Regarding works published at that time, Euclidean geometry was present only for a few rare authors who made specific references to propositions in the work. This was the case of Ibn Haydur in *Tuhfat al-tullab*, (The Adornment of the Students)[56] and al-'Uqbani in his commentary on the *Talkhîs* of Ibn al-Banna.[57] It was also present to the extent that these same authors demonstrated the propositions presented and the justifications of the constructions.[58] But most of the textbooks and in particular those concerning the geometry of measurement lack demonstrations.

In algebra, only two Maghrebian writings continue to support the teaching of this discipline: the poem by Ibn al-Yasamin and the *Kitab al-usul* [Book of Foundations] by Ibn al-Banna.[59] The two eastern books, *Mukhtasar* by al- Khwarizmi and *Kamil* by Ab Kamil that had nonetheless been circulated in Andalus and the Maghreb until the 13th century, are mentioned neither in biobibliographical sources nor in the mathematical writings of this time that have survived. We come to the same conclusion regarding the algebra books, written in Andalus, that are explicitly mentioned by Ibn Khaldun in his *Muqaddima* (Prolegomena).[60]

In the science of computation, it is the *Talkhîs* (*Summary of the Operations of Computation* by Ibn al-Banna), which is the reference work for teachers in Tlemcen.[61] In addition to the teaching of its contents, two eminent professors, al-'Uqbani and al-Habbak, have written two books about it in the form of commentaries.[62] A third teacher, Ibn Marzuq al-Hafid (d. 842/1438), even edited the work in the form of a poem so that its contents could be learned by heart.

Even stripped of the theoretical components, the mathematical activities of that time testify to the status of this discipline in the educated circles of the cities and its place in the training of the elite, especially jurists and theologians. This phenomenon seems to have been the consequence of the intervention of the States of the Maghreb in the training of Maghreb's administrators. The most significant initiatives in this field were those related to the construction of Madrasas and the financing of their activities by local authorities. The pioneers in this effort were the Merinids (1269–1465), who founded a number of institutions of this kind in cities in outer Maghreb.[63] The first motivations of the regional authorities in the creation of the madrasas were of a political and ideological nature. In the first place, to form the administrators who were to serve the State and secondly to strengthen the orthodoxy by controlling the training of future theologians whose main mission was to combat anything seen as a dangerous deviation. This intrusion of the State in the training of the future elites was not accepted unanimously by public opinion at the time. The process was even deemed harmful by some intellectuals who opposed it. Among them was the Tlemcen mathematician al-Abili (d. 1308) who spoke in these terms about the policies of the madrasas: "the proliferation of publications has distorted Science and the construction of madrasas made it disappear . . . because students are attracted to what is provided with stipends. Thus, we accept in <the madrasas> those that the people in power designate for teaching or those who agree to enter in their service. In fact, they have turned away from real men of science, those who we don't solicit or who, when they are solicited, don't answer, don't offer <as an answer> what we ask someone else."[64]

Mathematics in Central Maghreb During the Ottoman Phase: 16th to 19th Centuries

The Ottoman period of central Maghreb continues to cause problems for historians, particularly when it comes to defining the exact status of this province in relation to the central authority and the consequences of this for the analysis of the events and activities that distinguish this phase of the history of Algeria. Some University scholars consider this Ottoman presence as an outright occupation of an independent country by a foreign power.[65] For other scholars it is the continuation of the slow process of forming the modern State of Algeria under the protection of Ottoman oversight (against European attacks).[66]

In the scientific field, and particularly for mathematics and astronomy, no lapses in development occurred during this long period. Quantitatively, there are 50 well-known authors who published, after the 15th century, on one or other of these two disciplines. Qualitatively, the centuries of Ottoman domination of Algeria were part of the stagnation process observed in the previous two centuries. But it should be noted that this slowdown, then decline of scientific dynamism concerned, to varying degrees, affected the whole Muslim world, since it occurred also in the outer Maghreb which had remained outside the Ottoman hegemony. Also, we must add that even Istanbul's central government's measures to modernize scientific and technical activities did not change this situation. This modernization had some effect in Egypt and Ifriqiya but it had no impact on science education in central Maghreb.

As a province of the Ottoman Empire, this region had essentially military and financial links with the central government. It therefore retained a degree of autonomy at economic and political levels, in particular. Indeed, the majority of the population lived in rural areas in tribal structures (sedentary or nomadic) with essentially religious cultural activities coordinated by more or less powerful brotherhoods. As for the cities of the region, and in particular those that had known scientific activities during the preceding centuries, they managed to preserve a certain level of their scientific and cultural activities, in particular, thanks to the links they maintained with the intellectual centers of the other two regions of the Maghreb and in Egypt. They also benefited from the last contribution from al-Andalus after the expulsion of Muslims living there in 1609 by the Spanish King Philippe III.[67]

In the mathematical field, several dozen names of teachers or authors from central Maghreb have been cited. Their areas of interest were those of the previous century: practical geometry, science of computation (incorporating the basic elements of algebra), magic squares (for astrological purposes) and science of inheritances. The practical astronomy topics that continue to be taught are those supposed to facilitate the practice of religion. For example, the moments of daily prayers, the direction of Mecca, and the visibility of the Crescent Moon. In the wake of these teachings, there was the continuing invention and manufacture of astronomical instruments providing solutions to the first two problems.

Regarding pedagogical tools that accompany the teaching of these different subjects, there is a continuation and strengthening of the trends that appeared in the previous two centuries. In addition to traditional books on elementary geometry, Indian computation and distribution of inheritances, three forms of writings that are part of the pedagogy of repetition and memorization appear. The first form, the least widespread, is summaries or abstracts. The second is commentaries, in the form of independent publications or marginal glosses accompanying the commented text. This tradition began in the 14th century, as an extension of Ibn al-Banna's oral and written teachings. It continued until the beginning of the 19th century, as confirmed by the voluminous comments made in central Maghreb, by Shaykh Tfayyash (d. 1914) on the content of the *Kashf al-asrar 'an 'ilm huruf al-ghubar* [Unveiling the secrets of the science of dust figures] by al-Qalasadi."[68]

The third form is the versification of scientific material. The best-known mathematical poems produced in the 16th century in central Maghreb are those of al-Wansharsi (d. 1549) who set *Talkhîs* of Ibn al-Banna to verse and al-Akhdari (d. 1576). He directly edited in the form of a long poem a teaching manual entitled *al-Durra al-bayda'* (The white pearl). It deals with the two mathematical disciplines still taught in his time: the science of computation and the science of the distribution of inheritances.[69] He also put a few chapters of applied astronomy into verse under the title *al-Siraj fi 'ilm al-falak* (The lamp on astronomy). The contents of these two writings illustrate the trend to lower the level as has already been indicated. Indeed, if we consider the first poem as an example, we notice that its content is an abridged version of what had already been a summary in the 13th century, in the manual by Ibn al-Banna. This trend will moreover continue as is shown by the wide dissemination of al-Akhdari's first poem after the 16th century though the many copies that were made (many of them preserved in libraries of the Maghreb). And, above all, although the number of works published from the end of the 16th century until the 18th century that were dedicated to the explanation of its content.[70] Furthermore, the success of this poem continued in the 19th century, thanks to new commentaries but, above all, thanks to the advent of lithography which gave them an even wider distribution. This explains their presence until the early 20th century in the mathematics curriculum of the famous University of Theology in Tunis, the Zaytuna.[71] The poem even traveled to sub-Saharan Africa where it remained a reference until recent decades. It can be read in the manual on computation by the Malian Ahmad Baba al-Arawani (d. 1997).[72]

A few words are necessary on a topic that has not yet been researched. It concerns the languages used in central Maghreb in the teaching and the publication of mathematics during the Ottoman period. This question hasn't been asked until now because, to our knowledge, all mathematicians known in the Muslim West taught and wrote their works in Arabic during the period between the 9th and 15th centuries, with the exception of a few Jewish or Christian authors in al-Andalus, such as Abraham Bar Hiya, Abraham Ibn Ezra and the anonymous Christian author of the *Liber Mahameleth*.[73]

Some of these mathematicians were Berber and spoke *Tamazight*, i.e. the language of the inhabitants of the Maghreb before the Muslim conquest in the 7th century. But they wrote their works in Arabic. However, if we believe the Arab sources themselves, *tamazight* was used relatively early (between the 8th and the 9th century), in particular on the occasion of the first translation or adaptation of the Koran in that language.[74] In the 12th century, the Berber Almohad dynasty used this language to write texts popularizing their ideology, as did Ibn Tumart (d. 1130), the historic leader of this movement.[75] In the 16th century, religious books were translated into *Tamazight*. As for scientific texts and in particular, those written in that language with Arabic characters, they existed but they do not seem to have created a tradition of autonomous scientific teaching.[76]

The same question concerns the Turkish language (Osmanli) and its role in scientific activities in central Maghreb during the Ottoman period. Its use is confirmed, in particular by Ali Ibn Hamza (d. about 1611), a mathematician born in the city of Algiers. His book *Tufhat al-a'adad li dhawi al-rushd wa l-sadad* (The adornment of numbers for those who are gifted with reason and common sense) was written in this language.[77] But this mathematician lived mainly in Anatolia. Therefore, he is not representative of the group of practitioners of mathematics who taught in the cities of central Maghreb. The majority of them learned and taught mathematics in Arabic. This is the case for al-Wansharisi, al-Akhdari and Ibn Hamadush (d. about 1782), to name only the best known from this long period.[78]

Mathematics in Algeria's Colonial Period

In Algeria, the colonial power that was established gradually from 1830 and lasted until 1962 found educational institutions of different levels financed by the "waqf" system. As examples, at the beginning of the 19th century, there were, in Algiers, about 100 primary schools and four senior colleges (for a population of less than 20,000 inhabitants). In Tlemcen, there were two madrasas and fifty primary schools. In Constantine, there were ninety primary schools and seven senior colleges. The colonial administration was of course aware of this cultural reality, as confirmed by the testimony of General Daumas, "Primary education was much more prevalent in Algeria than generally believed. Our relationship with the local population of the three provinces has revealed that the average level of the male individuals who were literate was at least equal to the average level shown by departmental statistics in our country regions."[79]

The *waqf* system managed the "bequests" i.e. donations from private people in the form of money, buildings, cultivable land, shops or factories. These inalienable properties that accumulated for centuries were protected by legal documents.[80] The authorities quickly decided to close some teaching institutions. As for the others, especially those that were attached to *zawiyas* (Islamic educational institutions), the confiscation of the *waqf* possessions that had been allocated to them and the privatization of their lands resulted in their demise. Their managers could no longer pay the teachers, feed the students and maintain the school buildings. Another consequence of this decision was the exodus of a large number of teachers and educated people in general to neighboring countries and sometimes even to the Muslim East. These administrative decisions combined with the violence of the military occupation resulted in the disappearance of many of the cultural achievements of the previous centuries by the destruction of existing educational structures, the marginalization of classical Arabic, the asphyxiation of the surviving cultural and scientific production and also the scattering of archives and manuscripts. Throughout the 19th century, these policies led to the spread of illiteracy and obscurantism and at the same time the development of the culture of the *jihad* (holy war).[81]

After the consolidation of colonial power, new institutions emerged. In a first phase, the training they offered was intended almost exclusively for children of the settlers and the officials who ran the country.[82] With the marginalization of the Arabic language, French gradually became the language of science.[83] This benefited mathematics thanks to what the new educational system brought in the form of new programs, pedagogies and supervision. But in the first decades, only a few rare indigenous students benefited from this training, for three reasons: the insufficient number of schools, the selective policy which gave priority to the families of settlers[84] and the refusal of a large cross-section of the Algerian population to send their children to the new schools in protest to the occupation. This attitude evolved over time. But until the first decades of the 20th century, the French schools remained inaccessible to the majority of indigenous children.

What has just been said applies only to primary and secondary education. For higher education, it wasn't until 1879 that a "Preparatory School for the Teaching of Sciences" was established. Among the courses offered, there were higher mathematics courses that did not lead to a specialized diploma equivalent to a bachelor's degree.[85] In 1909, the colonial administration created the University of Algiers with a "Faculty of Science" replacing the "Preparatory School." This University also accepted students from Tunisia and Morocco (which had become French protectorates in 1881 and 1912 respectively). Attendance at this establishment by indigenous students was very low in proportion to the number of inhabitants. During the first twenty years of operation of this institution, the attendance rates of indigenous students did not exceed 5% of the student body. There were thirty Muslim students in 1914, forty-seven in 1920 and about a hundred at the end of the 1930s.[86] These numbers evolved slowly during the thirty years that preceded the country's independence.

Additional to the institutions similar to those in France, in 1850 the colonial administration created secondary schools, called *médersas*, for the bilingual training (French and Arabic) of an indigenous elite of a particular profile. Originally, the members of this elite group were not trained to become University students. They had to meet a recruitment requirement linked to the specific statute of Algerians living in the French State as persons to whom Muslim Law applied. Three *médersas* had been created for the whole country: in Constantine, Tlemcen and Medea. The latter was replaced by one in Algiers in 1859.[87] Students were often from modest social strata which had virtually no access to the other institutions. Their training would prepare them to become Arabic language teachers or members of the judicial authority in charge of indigenous affairs. From the middle of the 20th century under pressure from political events, the French administration decided to align the programs in French in those institutions with those of the French lycées. But it kept their specific characteristic, namely, on completing their education, students received a diploma equivalent to a bachelor's degree in Islamic law. This reform enabled the promotion of new indigenous graduates able to pursue academic studies and to study, in particular, a mathematical curriculum up to a bachelor's degree.

But the situation was not the same for all the country's inhabitants. If we limit our observations to the activities of the French minority, we note that the status of the sciences, especially mathematics, improved continually. One of the signs of this progress was the emergence of scholarly societies. As the status of the colony evolved, with the creation of three French departments in regions of Algeria, increased financial means were granted to the colonial administration to establish new university institutions. The strategic goal of this new orientation was to make the French presence more profitable by optimizing the country's economic output. One consequence of this policy was that between 1840 and 1909, a dozen scholarly societies emerged relating to the most important and most dynamic fields of activity, such as agriculture, medicine and geography. In the context of that time, the development of mathematics and related disciplines did not appear as a priority for the country. These different fields were therefore grouped with a "scholarly society of science" in 1863. But given the very low number of Algerian students attending the University of Algiers, this scholarly society remained exclusively composed of scientists of European origin for a long time.

We need to mention one more aspect of colonization. Though it is not of a scientific nature, it has had great importance after the country's independence. It concerns the cultural dimension of the teaching programs imported from France and introduced into the various training curriculums in institutions that operated during Algeria's colonial period. In their scientific training, the few thousands of Algerians who had the privilege of attending French schools never had the opportunity to learn about their country's scientific history. Even the mathematical knowledge that had been produced in the context of Islamic civilization and which formed part of the French educational programs was taught without historical and cultural dimensions. The situation was identical in secondary schools, especially in the *médersas*. The teaching there in Arabic was strictly limited to the literary and legal disciplines. After seven years of schooling, the student left the lycée without knowing anything of the political, cultural or the scientific history of his own country.

It seemed important to mention this last aspect of colonial education because of the delaying effects that have been observed during the first decades of the country's independence. And even today the men and women responsible for the teaching of the sciences, the development of the curriculums and the training of teachers are not insensitive to these historical and cultural aspects that we have just referred to.[88]

Mathematics During Algeria's Independence

In spite of 130 years of colonization by people and seven and half years of war, the first leaders of independent Algeria did not erase all the French heritage in matters of education and training. The same school system that functioned during the colonial period even survived until the middle of the 1970s, with only a few modifications. Until that date, the developments in mathematics were similar to those underway in France and concerned only the scientific content of the discipline. They did not yet express any of the cultural schisms that emerged after independence concerning the language of instruction.

The first phase of the country's educational policy was "quantitative" because it had to meet the very strong demand for schooling.[89] "Qualitative" aspects intervened with the launch of the first reform in 1976. This reform instituted "the fundamental school" lasting nine years and making Arabic the only language for teaching science subjects in primary and secondary schools. But the process of Arabization was not truly launched until the end of the 1970s and then only in a gradual manner. This is the reason it was not until 1993 that the first secondary education graduates completed all of their mathematical training in Arabic.

On the political front, the Arabization of primary and secondary education appeared very quickly as an instrument to consolidate the power of the single party by enlarging its social base. Pedagogically, it was a poorly controlled operation, both in its conception and in its implementation. It turned itself into a war machine against the French language, without achieving the objectives that were to make Arabic a powerful tool for scientific and cultural development. But the aspect that had the gravest consequences concerned the supervision of this Arabization.

In practice Arabization was entrusted to teachers, most of whom had not sufficiently mastered the fundamentals and the cultural elements of this language to be able to teach them properly (grammar, linguistics, literature, poetry). Then, when the increase in the learners required the recruitment of a still larger number of teachers, they turned to some countries in the Near East (Egypt, Iraq, Syria, Palestine) to supply contingents of teachers who were often badly prepared pedagogically for this new experience. On top of that, a significant proportion of these new recruits had ideological profiles inconsistent with both the global orientation of the Algerian State and the cultural traditions of the majority of its people. As an immediate consequence, there developed an ideological aspect of teaching in Arabic. Politically directed religious discourse inspired by the teacher's beliefs began to infiltrate the content of the teaching (even when it was scientific) and sometimes even to replace it when the topic allowed it. The situation got worse over time, especially since the majority of parents had not mastered the Arabic that was taught at school and so had no means of controlling the content of what was being taught. Moreover, the monopoly imposed by the single party did not allow the existence of associations that could react to these types of deviations.

This dynamic of a politically applied Arabization produced cultural and ideological antagonisms within the family of teachers, creating an actual divide between the education world and the University world. In this context of ideological effervescence, pedagogy didn't escape controversy, as demonstrated by the "battle of mathematical symbolism" which, for some time, divided the teaching staff in primary and secondary schools. More for cultural than pedagogical reasons, some teachers had rejected the Latin symbolism in use worldwide. They substituted an exclusively Arab symbolism that moreover had to be created and which was used for some time. The situation continued until the mid-2000s when we saw the return of the international symbolism at all levels of the teaching of mathematics throughout the country.[90]

From 1992 to 1994, decision makers of the time took advantage of the unification of three ministries (Education, Higher Education and Research) to develop a reform project involving the entire education system. The designers of the project had planned a special component for mathematics. But the political situation of this period, with many exacerbated ideological conflicts and violence of all kinds prevented the implementation of this reform. However, some of its provisions saw the light of day. Among them, there were new standards for the recruitment of teachers graduating from University (in order to raise the scientific level), the reform of teacher training institutes and the reintroduction of history as well as foreign languages in the examinations of the scientific sections of the baccalaureate.

In 2000, a 160-member national commission was created to develop a new reform. After lengthy debates that lasted over a year and during which all the cultural, political and ideological social sensibilities clashed, a consensus emerged to gradually reform the educational system. In this reform, which began to be implemented in 2003, public attitudes were in favor of the more effective teaching of mathematics. First, there was a return to ten years of pre-secondary education instead of nine, thus giving more time for assimilation of concepts and mathematical tools in the primary curriculum. Secondly, there was the decision to open the textbook market to the private sector with the aim of improving their quality. In mathematics, it was decided to gradually introduce two new topics in the school curriculum: probability and statistics.

But, considering the diverse positions and points of view that were represented on the national commission, a number of problems were not addressed, or when they were, there was no consensus on the recommended solutions. The first was bilingualism. In addition to its cultural and ideological dimensions (that never allowed the establishment of a calm debate on the topic), there were pedagogical aspects concerning the mastery of the Arabic language and its necessary "cohabitation" with foreign languages. In this sphere, we need to specify that the languages that are taught are in the same situation insofar as, at the end of their studies, a large number of students are not very fluent in either classical Arabic, or French or the other foreign languages introduced after independence. For years, this situation that students experienced affected their ability to understand and assimilate the mathematical or generally scientific discourse of their teachers. The second problem concerns the cultural dimensions of the scientific subjects taught. The first part of this modest study has shown the importance of the scientific heritage of the Maghreb as having participated in the development and teaching of mathematics in relation to the most practiced disciplines at that time, such as geometry, algebra, the science of computation and the theory of numbers. However, none of the reforms implemented in Algeria contain serious reflection on the place and role in of history of this discipline and its role in the teaching of its concepts, its methods and its tools. The situation is not very different in the other countries of the Maghreb.

The third and final problem facing the leaders who succeeded one another at the head of the Ministry of National Education is the level of competence of the supervisors (teachers and inspectors) in charge of the teaching of mathematics. Apart from official speeches and statistics about the different categories of teachers and staff numbers, there are few qualitative analyses on the history of their recruitment, their starting level, the process of their training, their internal promotion criteria and their behavior in front of students.

Mathematics at Universities

The policies of the colonial administration in Algeria did not allow many students with a bachelor's degree to proceed to the University. Indeed, like the image of the education sector, higher education during colonization reflected at every level the unequal treatment of the two communities that cohabitated the country. It was not until World War I that the number of Muslim students reached 1% of the total of enrolled students. And less than seven years prior to independence, Algerians enrolled in any of the three existing University structures (Algiers, Oran and Constantine) accounted for slightly less than 13% of all students (whereas their people comprised 90% of the total inhabitants of the country). As discussed in the preceding chapter, the delayed initiatives taken by the colonial administration did not permit a significant increase in the number of students with bachelor's degrees able to take higher studies in mathematics or closely related scientific disciplines like physics and mechanics. At the start of the new 1962 academic year, which was the first year of the country's independence, the number of mathematics students was so small that the lecture theaters were too large for students to feel welcome. This imbalance also existed in the supervisors. In the aftermath of independence, there were no professors of Algerian origin and the number of teachers of lower rank was so small that an appeal was made to the solidarity of French academics who had campaigned for the independence of the country. This appeal was heard, and a number of senior teachers agreed to accompany the Algerian University's first steps, contributing greatly especially in the Faculties of science.

The first serious problem that occurred at institutions of higher learning came into the open at the beginning of the 1993–1994 academic year. That was the first year that students with a scientific education that was completely in Arabic joined the ranks of the University. Their mastery of French scientific terminology was slim to non-existent. Serious consequences resulted for their ability to understand the scientific discourse of their teachers and for the level reached by the majority of them at the end of the academic year. This situation worsened for the next cohorts of students because the gap continued widening between the secondary schools and higher education. On the one hand, owing to the influence of cultural and ideological pressure that favored total Arabization, there was a steady deterioration of the teaching of the French language before admission to universities. On the other hand, the training exclusively in Arabic of the future teachers of the scientific disciplines had long created a gap between these teachers and university administrators who had the capacity to propose updated knowledge in these disciplines and teaching methods adapted for teachers in secondary schools.[91]

On the research front, the accumulated deficit during the second period of colonization limited the rapid formation of a community of high-level mathematicians. After independence, the Department of Mathematics at the University of Algiers had a majority of foreign teachers. It was not until the middle of the 1980s that about fifteen holders of doctorates in mathematics were Algerian.[92] Then, along with the quantitative development of the teaching body in this discipline, it seemed necessary to create the first national scholarly society responsible for bringing together all people active in the field (teachers of different university levels and researchers). This could be done only when the right to form an association (without the oversight of the single party) was in reality allowed. That is why, at the end of the *Premier Séminaire National des Mathématiciens Algériens*, (the First National Seminar of Algerian Mathematicians), which took place in 1977, the *Société Algérienne de Mathématiques* (Algerian Mathematical Society) came into existence. Initially, there was no political reason or ideological motive for the creation of this body. On the contrary, the promoters of this project were motivated only by professional concerns. But the single political party was still powerful, and its leaders did not see any initiatives that challenged its monopoly in good light. It is probably the main reason why the first attempt failed. The second attempt took place after the 1988 protests, which were directed against the established power and its single party. It led to the creation of *l'Association Mathématique Algérienne* (Algerian

Mathematical Association), that four years later launched the *Revue Maghrébine de Mathématiques*, (Maghrebian Mathematical Review). But the political and economic situation in the country during the 1990s did not favor the growth of the association. Indeed, in this period, Algerian society lived through multiple crises brought on by the collapse of the prices of oil and gas resulting in a sharp reduction of income for the State, rejection of the political governance by a majority or the population at the 1990 municipal elections and the 1991 legislative elections, the emergence of new political movements of a cultural or religious nature, the appearance and development of terrorism as a new political instrument. This situation resulted in discontinuation of the activities of the association, its dissolution in 2009 and the establishment of the *Société Mathématique Algérienne* (Algerian Mathematical Society), which is still existent today.

By Way of Conclusion

In recent years, Algerian universities (which today number 50) have continued to produce students with majors in mathematics. But the results are less than what could be expected from the country's financial investment. The number of graduates with sufficient credentials is not enough to both renew the Faculty and contribute to the research sector. At the same time, as has been seen for decades, despite a number of initiatives to curb the phenomenon, there is a sharp loss of the best graduates, by way of the "brain drain" to other countries. Of course, the community of mathematicians suffers from this hemorrhaging, similarly to other sectors. Another phenomenon troubling policy makers for the near future, is the trend of the best secondary school students to abandon mathematics studies in favor of others deemed more "profitable." Measures have been taken to slow down and then stop this phenomenon. One of them was the opening at the beginning of the 2012–2013 school year in Algiers, of an "institution of excellence." It brings together students from the whole country who obtained the best results in mathematics in the "Brevet de l'Enseignement Elémentaire" examination to prepare them for entry into the mathematical courses at University. The "Olympics" have also been revived to motivate students to study mathematics from an early age and to discover talents. At the same time, discussions are taking place at different levels to circumvent an even more serious phenomenon, because it is educational in nature. It concerns the way the teacher transmits mathematical knowledge, or know-how, and how the student understands, assimilates and uses this.

In fact, there is another troubling trend that is also present in the educational systems of other countries. It appeared a few years ago and continues to grow. It is learning mathematics by memorization of the processes for solving given problems, rather than the assimilation of investigative methods by reflection, analysis and the mastery of the mathematical tools and processes by use. The first part of this study showed that the Ottoman period of central Maghreb experienced the same phenomenon with its negative consequences for the vitality of the sciences that were still performed, especially astronomy and mathematics. This is a lesson from history that policymakers should contemplate.

References/Notes

[1] Ch. A. Julien: *Histoire de l'Afrique du nord.* Vol. I: *Des origines à la conquête arabe*, Paris, Payot, 1975: A. Laroui: *L'Histoire du Maghreb, un essai de synthèse*, Paris, Maspéro, 1970, pp. 21–76.

[2] P. Gerdes: *Geometry from Africa: Mathematical and educational explorations*, Washington, DC, The Mathematical Association of America, 1999.

[3] Ch. Gillispie (ed.): *Dictionary of Scientific Biography*, New York, Scribner's Sons, 1981, vol. 1, pp. 333–338.

[4] Ch. A. Julien: *Histoire de l'Afrique du Nord*, vol. I, op. cit., pp. 256–279.

[5] Y. Thébert: Royaumes numides et hellénisme, *Afrique & Histoire*, no. 1, vol. 3 (2005), pp. 29–37: H. Kadra-Hadjadji: *Massinissa le grand africain*, Paris, Editions Karthala, 2103, pp. 148–149.

[6] A. Djebbar & Y. Guergour: La numération *rumi* dans des écrits mathématiques d'al-Andalus et du Maghreb avec l'édition d'une épître d'Ibn al-Banna, *Suhayl*, no. 12 (2013), partie arabe, pp. 7–52.

[7] A. Djebbar: On mathematical activities in North Africa since the 9th century. First part: Mathematics in medieval Maghreb. *AMUCHMA Newsletter*, no. 15 (1995), pp. 4–45.

[8] F. G. Fitz: *La batalla de Las Navas de Tolosa: el impacto de un acontecimiento extraordinario*, Actes du Colloque International « *Las Navas de Tolosa 1212–2012,*

Miradas Cruzadas » (Jaen, 9–12 avril 2012), P. Cressier & V. Salvatiera (ed.), Jaen, Universidad de Jaen, 2014, pp. 11–36.

[9] Ch. A. Julien: *Histoire de l'Afrique du Nord*, vol. II: *De la conquête arabe à 1830*, Paris, Payot, 1969, pp. 38–40, 66–75, 154–193: A. Laroui: *L'Histoire du Maghreb, un essai de synthèse*, op. cit., pp. 86–96, 186–206.

[10] A. Djebbar: *Les activités mathématiques dans les villes du Maghreb Central (IXe-XVIe s.)*, Actes du 3e Colloque Maghrébin sur l'Histoire des Mathématiques Arabes (Tipaza, 2–4 Décembre 1990), Alger, Office des Presse Universitaires, pp. 73–115.

[11] A. Djebbar: *Quelques éléments nouveaux sur les activités mathématiques arabes dans le Maghreb oriental (IXe-XVIe siècles)*, Actes du 2e colloque maghrébin sur l'Histoire des mathématiques arabes (Tunis, 1–3 décembre 1988), Tunis, I.S.E.F.C.-A.T.S.M., 1990, pp. 55–61.

[12] Abu Zakariya': *Kitab siyar al-a'imma wa akhbarihim* [The book on the life of the Imams and their actions], I. Al-'Arabi (ed.), Alger, Bibliothèque Nationale, 1979, p. 35.

[13] Ibn Saghir: *Kitab al-tarikh* [History book]. In Motylinswki (ed. & trad): *Chronique d'Ibn Saghir sur les imams rostémides de Tahert*, Actes du XIVe Congrès des Orientalistes (Alger, 1905), 3e partie, Paris, Ernest Leroux éditeur, 1908, texte arabe, p. 26.

[14] Al-Shammakhi: Tabaqat al-Ibadiya [Ibadi Categories]. Cité par M. I. 'Abd Al-Raziq: *Al-Khawarij fi bilad al-Maghrib* [The Kharijites in the Maghreb], Casablanca, Dar al-thaqafa, 1976, p. 15.

[15] Abu Zakariya': *Kitab siyar al-a'imma wa akhbarihim*, op. cit., p. 89.

[16] Op. cit. p. 132.

[17] Ibn bushkuwal: *Kitab al-sila* [Book link] Le Caire, Al-Dar al-misriya li l-ta'lif wa l-tarjama, 1966, vol.1, p. 298, no. 656.

[18] A. Djebbar: *Quelques éléments nouveaux sur les activités mathématiques arabes dans le Maghreb oriental (IXe-XVIe siècles)*, op. cit., pp. 63–64.

[19] Al-Ghubrini: *'Unwan al-diraya fi man 'urifa min al-'ulama' fi l-mi'a al-sabi'a fi Bijaya* [Ornament of knowledge on those known scholars of Bejaïa in the seventh century AH] 'A. Nuwayhid (ed.), Beyrouth, Dar al-afaq al-jadida, 1979.

[20] Ibn al-'Imad: Shadharat al-dhahab [The gold glitters] Le Caire, 1949, vol. IV, p. 158.

[21] 'A. Ibn Khaldun: *Kitab al-'ibar* [The Book of sentences] Beyrouth, Dar al-kitab al-lubnani-Maktabat al-madrasa, 1983, vol. XI, p. 350.

[22] Al-Warjalani: *Al-Dalil wa l-burhan* [The argument and the proof], S. Al-Harithi (ed.), Mascate, Ministère du Patrimoine National et de la Culture, 1983, pp. 107–145.

[23] Ibn Farhun: *Al-Dibaj al-mudhahhab fi ma'rifat a'yan 'ulama' al-madhhab* [The gold brocade for the knowledge of eminent scholars of the rite], Beyrouth, Dar al-kutub al-'ilmiya, non datée, pp. 90–91.

[24] Al-Ghubrini: *'Unwan al-diraya*, op. cit., pp. 57–61.

[25] Op. cit., pp. 495–496

[26] A. Djebbar: *Enseignement et recherche mathématiques dans le Maghreb des XIIIe-XIVe siècles*. Paris, Université de Paris-Sud, Publications Mathématiques d'Orsay, 1980,

no. 81–02, p. 10: M. Zerrouki: Abu l-Qasim al-Qurashi, (m. 580/1184), savant en mathématiques et en science des héritages, *Cahier du Séminaire Ibn al-Haytham*, Alger, 1995, no. 5, pp. 10–19. (en arabe).

[27] A. Djebbar: *L'algèbre arabe, genèse d'un art*, Paris, Adapt-Vuibert, 2005, pp. 79–82.

[28] E. Laabid: *Ibn Safwan al-Malaqi (m. 1362) et sa contribution dans la tradition mathématique des héritages*, Actes du 10e Colloque maghrébin sur l'Histoire des mathématiques arabes (Tunis, 29-31 mai 2010), Tunis, Publications de l'Association Tunisienne des Sciences Mathématiques, 2011, pp. 198–210.

[29] 'A. Ibn khaldun: *Kitab al-'ibar*, op. cit., vol. II, p. 773: A. B. Al-Tinbukti: *Nayl al-ibtihaj bi tatriz al-dibaj* [the acquisition of joy with brocade embroidery], Dar al-kutub al-'ilmiya, non datée, pp. 344–345.

[30] Ibn al-Khatib: *Al-Ihata fi akhbar Gharnata* [The comprehensive book on Granada's chronicle], M. A. 'Inan, Le Caire, Maktabat al-Khanji, 1975, vol. III, pp. 324–330.

[31] Ibn al-Qadi: *Durrat al-hijal fi asma' al-rijal* [The Pearl of matrimonial beds on the names of famous men], Le Caire, Dar al-turath-Tunis, al-Maktaba al-'atiqa, 1970, vol. III, pp. 187–188, no. 1172.

[32] A. Djebbar: *L'analyse combinatoire au Maghreb: l'exemple d'Ibn Mun'im (XIIe-XIIIe siècles)*. Paris, Université Paris-Sud, Publications Mathématiques d'Orsay, 1985, no. 85-01: A. Djebbar: Figurate Numbers in the Mathematical Tradition of Andalus and the Maghrib, *Suhayl*, Barcelone, no. 1 (2000), pp. 57–70.

[33] D. A. King: *An overview of the sources for the history of Astronomy in the medieval Maghrib*, Actes du 2e Colloque Maghrébin sur l'Histoire des Mathématiques Arabes (Tunis, 1-3 décembre 1988), Tunis, I.S.E.F.C.-A.T.S.M., 1990, pp. 128–130: A. Mestres: *Maghribi Astronomy in the 13th Century: A Description of Manuscript Hyderabad Andra Pradesh State Library 298*. In J. Casurellas & J. Samso (ed.): *From Baghdad to Barcelona, Studies in the Islamic Exact Sciences in Honour of Prof. Juan Vernet*. Barcelona, Anuari de Filologia XIX - Instituto Millas Vallicrosa, 1996, pp. 383–443.

[34] A. Djebbar & M. Aballagh: *La vie et l'œuvre d'Ibn al-Banna: un essai biobibliographique*, Rabat, Université Mohamed V, Publications de la Faculté des Lettres et Sciences Humaines (en arabe), 2001, pp. 30–43.

[35] Op. cit., pp. 89–109.

[36] M. Souissi: *Talkhîs a'mal al-hisab* [The abridged book on computing operations], Tunis, Publications de l'Université de Tunis, 1969. p. 111.

[37] 'A. Ibn Khaldun: *Al-Ta'rif bi Ibn Khaldun* [Autobiography of Ibn Khaldun], Beyrouth, Dar al-kitab al-lubnani-Le Caire, Dar al-kitab al-misri, 1979, p. 8.

[38] A. Harbili: *L'enseignement des mathématiques à Tlemcen au XIVe siècle à travers le commentaire d'al-'Uqbani (m. 1408)*, Magister en Histoire des mathématiques, Alger, E.N.S., 1997.

[39] Ibn al-Khatib: *Al-Ihata*, op. cit., vol. III, p. 507.

[40] Ibn al-Qadi: *Durrat al-hijal*, op. cit., vol. II, p. 38.

[41] Y. Guergour: *Al-A'mal al-riyadiya li Ibn Qunfudh al-Qasantini*, Mémoire de Magister en Histoire des mathématiques, Alger, Ecole Normale Supérieure, 1990.

[42] E. S. Kennedy & D. A. King: Indian Astronomy in Fourteenth Century Fez, the Versified Zij of al-Qusuntini, *Journal for the History of Arabic Science*, 6, 1982, pp. 3–45.

[43] J. Samso: Andalusian Astronomy in 14th Century Fez: al-Zij al-Muwafiq of Ibn 'Azzuz al-Qusantini, *Zeitschrift für Geschichte der Arabisch-Islamischen Wissenschaften* no. 11 (1997), pp. 73–110.

[44] D. A. King: *An overview of the sources for the history of Astronomy in the medieval Maghrib*, op. cit., pp. 125–157.

[45] 'A. Ibn Khaldun: *Kitab al-'ibar*, op. cit., vol. II, p. 774.

[46] Y. Ibn Khaldun: *Bughyat al-ruwwad* [The desire of precursors], A. Hadjiat (ed.), Alger, Bibliothèque Nationale, tome I, 1980, pp. 106–107, 119, 127, 207, 209: Ibn Maryam: *Al-Bustan* [The garden], Alger, Office des Publications Universitaires, 1986, pp. 33–34, 70, 92, 274–275.

[47] 'A. Ibn Khaldun: *Al-Ta'rif bi Ibn Khaldun*, op. cit., pp. 33–39, 64–65.

[48] E. Laabid: *Les problèmes d'héritage et de mathématiques au Maghreb des XIIe-XIVe siècles, essai de synthèse*, Actes du 7e Colloque maghrébin sur l'Histoire des mathématiques arabes (Marrakech, 30 mai-2 juin 2002), Marrakech, Al-Wataniya, vol. I, pp. 241–261.

[49] Op. cit., p. 31.

[50] Ibn Maryam: *Al-Bustan*, op. cit., p. 65.

[51] Y. Ibn Khaldun: *Bughyat al-ruwwad*, vol. II, Ms. Paris, B.N. 5031, p. 136: A. Hadjiat: *Abu Hammu Musa al-Ziyani, hayatuhu wa atharuhu* [Abu Hammu Musa al-Ziani, his life and work], Alger, S.N.E.D., 1974.

[52] Al-Qalasadi: *Rihla* [Travel Report], M. Abu l-Ajfan (ed.), Tunis, Société Tunisienne de diffusion, 1978, pp. 103–104.

[53] Op. cit., pp. 103–104.

[54] M. Baghli: *Madaris wa masajid Tilimsan qabla 'ahd al-ihtilal* [The madrasas and mosques of Tlemcen before colonization], Speech on the occasion of the seven hundredth anniversary of the opening of the madrasa al-Ya'qubiya, Tlemcen, November 14, 1993, p. 19.

[55] M. L. Khattabi: Risalatan fi 'ilm al-misaha li Ibn al-Raqqam wa Ibn al-Banna [Two epistles on the science of measurement of Ibn al-Raqqam and Ibn al-Banna], *Da'wat al-haqq* (Rabat), 1986, no. 256, pp. 39–47: M. L. Khattabi: Sharh al-Iksir fi 'ilm al-taksir li Abi 'Abdallah Ibn al-Qadi [Commentary on 'Elixir on the science of measurement' of Abu 'Abdallah Ibn al-Qadi], *Da'wat al-haqq*, no. 258, pp. 77–87.

[56] Ibn Haydur: *Tuhfat al-tullab* [The adornment of students] Ms. Rabat, Bibl. Al-Hasaniya, no. 252, ff. 29b, 31a, 71a, 133b, 134b.

[57] A. Harbili: *L'enseignement des mathématiques à Tlemcen au XIVe siècle*, op. cit.

[58] Op. cit.

[59] Al-Qalasadi: *Rihla*, op. cit., p. 101.

[60] 'A. Ibn Khaldun: *Kitab al-'ibar*, op. cit., vol. II, p. 898–899.

[61] Al-Qalasadi: *Rihla*, op. cit., p. 101: Ibn Maryam: *Al-Bustan*, op. cit., p. 276.

[62] A. Djebbar & M. Aballagh: *La vie et l'œuvre d'Ibn al-Banna*, op. cit.

[63] M. Kably: *Qadiyat al-madaris al-marriniya, mulahazat wa ta'mmulat* [The issue of Marinid madrasas, remarks and reflections]. In M. Kably: *Revue de la société et de la culture dans le Maroc médiéval*, Casablanca, Editions Toubkal, 1987, pp. 66–78.

[64] A. B. Al-Tinbukti: *Nayl al-ibtihaj*, Beyrouth, Dar al-kutub al-'ilmiyya, op. cit., p. 246.

[65] A. Sa'dallah: *Tarikh al-jaza'ir al-thaqafi min al-qarn al-'ashir ila al-qarn al-rabi' 'ashar al-hijri* [Cultural history of Algeria, from the tenth to the fourteenth century AH], Beyrouth, Dar al-Gharb al-islami, 1998, vol. 1, pp. 14–15.

[66] A. T. Al-Madani: *Harb ath-thalath mi'at sana bayna l-Jaza'ir wa Isbanya (1492-1792)* [The war of three hundred years between Algeria and Spain (1492-1792)], Alger, Société Nationale d'Édition et de Diffusion, 1976.

[67] L. Caradaillac (dir.): *Les Morisques et l'inquisition*, Paris, Publisud, 1990.

[68] A. Djebbar & M. Moyon: *Les sciences arabes en Afrique, Astronomie et mathématiques (IXe-XIXe siècles)*, Paris, Grandvaux-Vecmas, 2011, pp. 79–80.

[69] 'A. Al-Akhdari: *Al-Durra al-bayda* [La perle blanche], Le Caire, lithographie, 1891.

[70] A. Sa'dallah: *Tarikh al-jaza'ir al-thaqafi*, op. cit., vol. II, p. 405–406.

[71] M. Souissi: *Tadris al-riyadiyat bi l-'arabiya fi l-Maghrib al-'arabi wa khassatan bi Tunus fi l-qarn al-thalith 'ashar li l-Hijra* [The teaching of Arabic mathematics in the Arab Maghreb, especially in Tunisia in the thirteenth century and the first half of the fourteenth century AH], Actes du 3e Colloque Maghrébin sur l'Histoire des Mathématiques Arabes (Alger, 1–3 décembre 1990), Alger, Office des Publications Universitaire, 1998, partie arabe, p. 32.

[72] A. Djebbar & M. Moyon: *Les sciences arabes en Afrique*, op. cit., pp. 110, 144.

[73] A. Djebbar: *La circulation de l'algèbre arabe en Europe et son impact.* Actes du colloque international sur « *The Impact of Arabic Sources in Europe and Asia* » (Erlangen, 21–23 janvier 2014). In *Micrologus* XXIV, Florence, Sismel-Edizioni Galuzzo, 2016, pp. 103, 105, 107–108.

[74] Ch.-A. Julien: *Histoire de l'Afrique du Nord*, vol. II, op. cit., p. 39: A. Laroui: *L'histoire du Maghreb*, op. cit., p. 104.

[75] R. Bourouiba: *Ibn Tumart*, Alger, Société Algérienne d'Édition et de Diffusion, 1974, p. 103.

[76] A. Amahan: *Notes bibliographiques sur les manuscrits en langue tamazight écrits en caractères arabes*. In A. Ch. Binebine (ed.): *Le manuscrit arabe et la codicologie*, Rabat, Université Mohamed V, Publications de la Faculté des Lettres et Sciences Humaines, Série Colloques et Séminaires, no. 33, 1994, pp. 99–104.

[77] S. Zeki: *Athar baqiya* [The remaining vestiges]. Cité par Q. H. Tuqan: *Turath al-'Arab al-'ilmi fi l-riyadiyat wa l-falak* [The Arab Scientific Heritage in Mathematics and Astronomy], Beyrouth, Dar al-shuruq, 1963, pp. 470–471.

[78] A. Sa'dallah: *Tarikh al-jaza'ir al-thaqafi*, op. cit., vol. 2, pp. 425–436.

[79] Y. Turin: *Affrontements culturels dans l'Algérie coloniale*, Paris, Maspéro, 1971, p. 127.
[80] P. J. Bearman & Al. (ed.): *Encyclopaedia of Islam*, Leiden, Brill, 2002, vol. XI, pp. 59–99.
[81] Y. Turin: *Affrontements culturels dans l'Algérie coloniale*, op. cit., 1971, pp. 119, 134–135: Ch.-H. Churchill: *The Life of Abdel Kader, ex Sultan of the Arabs of Algeria*, London, Chapman & Hall, 1867.
[82] In 1912, only 4.7% of Muslim children were in school, with very few girls. And, in that proportion, many of the children who go to school were the sons of Muslim notables in the service of the colonial administration. In 1930, the rate did not exceed 9%. Ch.-R. Ageron: *Histoire de l'Algérie contemporaine*, Paris, Presses Universitaires de France, 1979, t. 2, p. 163: F. Colonna: *Instituteurs algériens 1883–1939*, Paris, Presses de la Fondation Nationale des Sciences Politiques, 1975, p. 50.
[83] Y. Turin: *Affrontements culturels dans l'Algérie coloniale*, op. cit., pp. 196–303.
[84] Of the 6500 primary classes in 1944, 5500 were occupied by 118,000 Europeans and about a thousand by only 108,000 Algerians. Ch.-R. Ageron: *Histoire de l'Algérie contemporaine*, op. cit., pp. 534–535.
[85] L. Paoli: L'enseignement supérieur à Alger, *Revue Africaine*, vol. 49 (1905), pp. 406–437.
[86] P. Singaravélou: L'enseignement supérieur colonial, un état des lieux, *Histoire de l'éducation*, no. 122 (2009), pp. 71–92.
[87] Ch. Janier: Les médersas algériennes de 1850 à 1960, *Mémoire Vive*, no. 46, 3e-4e trimestre (2010), pp. 2–35.
[88] A. Djebbar: Les scientifiques arabes face à leur patrimoine, *Maghreb-Machrek*, no. 105 (1984), pp. 48–64.
[89] The year before the country's independence, over 80% of school-age children (that is, between 6 and 15 years) were not enrolled in school. Ch.-R. Ageron: *Histoire de l'Algérie contemporaine*, op. cit., p. 533.
[90] M. Belabbès: *Al-Riyadiyat* [Mathematics], 1e année secondaire, filière commune «*Science et Technologie*», Alger, Office National des Publications Scolaires, 2005.
[91] Commission Nationale des Programmes: *Les programmes et les documents annexes, Mathématiques, 2e année du secondaire général et technologique*, Alger, Office National des Publications Scolaires, 2006.
[92] P. Gerdes: *African Doctorates in Mathematics: A catalogue*, 2007, Maputo, Research Centre for Mathematics, Culture and Education, 2009, pp. 31–32.

About the Author

Dr. Ahmed Djebbar is an Algerian born mathematician and historian. In the past, he was advisor of the Algerian President Mohamed Boudiaf. He served as the Minister of Education and Research in Algeria from July of 1992 to April of 1994. He is Professor Emeritus of History of Mathematics at the University of Science and Technology Lille. As a research scientist, Dr. Djebbar specializes in medieval Arab mathematics of al-Andalus, Maghreb and sub-Saharan Africa.

https://doi.org/10.1142/9789813146785_0003

Chapter III

EGYPT: Mathematics Education in Egypt

Ahmed Ghaleb
University of Cairo

Soha Abdelsattar
Teachers College, Columbia University

Ancient Egyptian Mathematics

Egypt has long been a site for mathematical study and progress. Egyptian mathematics dates back to at least 3000 BC, preserved in richly-inscribed walls and papyri (Eves, 1990). The solar calendar was invented by the Egyptians. The monuments form a tangible proof of the mathematical knowledge in Egypt. Among the most popular historical mathematical findings of the time was the Moscow papyrus, dating back to around 1850 BC, containing 25 mathematical problems. Another famous sample of mathematic works was the Rhind (or Ahmes) papyrus, dating back to approximately 1650 BC. This text contained 85 mathematical problems. All of the problems described in the Moscow and Rhind papyri are numerical, and most are practical in nature (Eves, 1990). The Moscow papyrus contains a numerical description of the correct formula for the volume of a frustum of a square pyramid. This is considered a remarkable accomplishment for its time, and has been described as the "greatest Egyptian pyramid" (Eves, 1990, p. 55).

The mathematics of ancient Egypt has therefore set a high standard for its people to maintain, establishing Egypt as a center of civilization and education. Recorded history on the details of teaching and learning of mathematics, however, did not begin until centuries later.

The Alexandria School

The Alexandria School developed and flourished in the Mediterranean city of Alexandria, Egypt during the Hellenistic and Roman periods for a few centuries, vaguely extending from 300 BC to 400 AD. The School witnessed the activity of some of the most celebrated mathematicians, such as Thales, Pythagoras, Euclid, Archimedes, Diophantus, Ptolemy, Heron the Alexandrine and Hypatia.

The Middle Ages

Egyptian mathematicians in the Middle Ages, a period corresponding to the apogee of the Moslem era, worked mostly on algebraic and geometrical problems. However, they had also inherited from their predecessors, of the Alexandria school, a large amount of knowledge in mastering mathematics for applications, especially in the fields of astronomy, navigation, mechanisms and war engines. They can thus be truly called the scientific heirs of celebrated Alexandrine ancestors. Under Moslem era, many Copts (Egyptian Christians) were capable of performing accountability works and these were encouraged to work for the new system to manage the resources of the country, including tax. The most prominent Egyptian mathematicians of that time were Abou Kamil Ibn Aslan Shuja (c. 850–c. 930) the predecessor of Fibonacci, who used irrational numbers and applied algebraic methods to geometrical problems, Abul' Hassan Ali Ibn Yunus (950–1009) whose astronomical applications of trigonometry are celebrated and Ibn Shehab the Cairene (c. 1446) in astronomy (2008, قدري طوقان).

During these centuries, a structured educational system had not yet been established in Egypt, nor in most parts of the world. Most young Egyptian pupils of the time relied on the "kuttab" system for early education, as this was the closest arrangement to organized schooling. The main purpose of this system was the learning of the Arabic language through memorizing the Quran, although students could still emerge illiterate (Heyworth-Dunne, 1939). Therefore, the general population were not, and could not be, introduced to areas of study such as mathematics.

Recent History

It is difficult to trace back global mathematical activity in Egypt before the 19th century. Certainly, there was a marginal interest in mathematics from the official theological institution represented by "Al-Azhar," driven mainly by religious purposes like calculating dates and establishing calendars, as well as elements of traditional logic. The college-mosque of Al-Azhar, which had been founded in Cairo in the year 972 AD ("Al-Azhar University," n.d.), was one of the most prestigious forms of higher education at the time. While about half of the curriculum was comprised of religious education, some topics in mathematics were also taught under a larger theme which was called *the rational sciences*. This area of study was a rather difficult jump for students who had no mathematics training after completing their kuttab education.

The aftermath of the Napoleonic campaign in Egypt (1799–1801) witnessed the rise of what can be named modern Egypt under the rule of Mohammed Ali Pascha, who followed the tradition of taking the Salahuddin Citadel in Cairo as his headquarters. To support Mohammed Ali's ambition and strategic views concerning Egypt's interest in the surrounding region, overall reform was necessary. Public schools were built in the different urban centers, where mathematics appeared for the first time among the curriculum subjects. These were financed and run by the governmental institutions, separate from the Al-Azhar umbrella.

Unprecedented efforts were undertaken in re-organizing the Egyptian army and the fleet, and modernizing the irrigation system in the country. Mathematics was desperately needed in both fields: to control the water of the Nile and for navigation and artillery purposes. Mathematics had become a sure way for fame and jobs. As time was the most precious factor, Mohammed Ali found that the fastest way to obtain the needed experts was to send selected youth abroad to complete formation in mathematics as well in other sciences. France was a preferred destination, which is fully understandable in light of the recent contact with the French.

Let us not forget that the famous French mathematician Joseph Fourier had accompanied Napoleon in the expedition to Egypt a few years earlier. Upon return from abroad, the young scientists would be literally imprisoned in the Citadel until they translated a book in their field of specialization into Arabic. The ruler would confer the honorary title of Pascha upon many of them.

Two important steps were to mark the progress of mathematics teaching in Egypt: the founding of the School of Engineering and that of the High School for Teachers.

The School of Engineering

For the first time in Egypt, a school of engineering (Madrasat Al-Mohandeskhana) was founded in 1816 based at the Citadel. It had two sections: irrigation and civil engineering. Mathematics was also taught, including algebra, geometry and trigonometry.

Among the prominent figures of that time was the renowned Egyptian mathematician and astronomer Mahmoud Hamdi (1815–1885), who acquired later on the appellation El-Falaki (the astronomer, in Arabic) in view of his extended scientific activities in astronomy. Hamdi was sent to France where he stayed for 9 years studying astronomy and mathematics. Another renowned Egyptian mathematician of that time is Mohamed Bayyumi Effendi. He spent nine years in Paris studying mathematics and engineering. Upon his return to Egypt, Bayyumi wrote and translated several works on algebra, trigonometry and mechanics (History of Sciences and Engineering Technology in Egypt in the 19th and 20th centuries).

Mahmoud El-Falaki Pascha
1815–1885

The School of Engineering was reshuffled in 1834 and was transferred to the quarter of Boulak (beau lac, in French), but shut its doors in 1854. Independently, a School for Irrigation was founded in the suburb of Kanater in 1858, nearby the emplacement of a dam. Another School for Civil Engineering in the Citadel was founded around the same time. However, both were closed in 1861.

In 1866, the old School of Engineering was revived, but in the quarter of Abbasseya. Only twenty years later was the curriculum of study in this school set to five years. Mathematics were taught intensively during the first two years. In 1925 this school became the Faculty of Engineering at the Egyptian University (Fouad I University since 1940, then Cairo University since 1953).

Ali Mubarak Pascha
1823–1893

A central figure in the modern reform in Egypt was Ali Mubarak. He graduated from the School of Engineering and was sent to France, where he spent five years studying mathematics, astronomy, chemistry and military geometry for artillery purposes. With the increasing number of schools in Egypt, there was a need to create a centralized administration for them. Ali Mubarak was chosen as administrator of school affairs. Later on, he was nominated Minister of Education and contributed largely in reforming the schools, with particular interest in the teaching of geometry. Ali Mubarak is considered the founder of Egypt's modern educational system.

The Higher School for Teachers

One of the major steps in the reform of education in Egypt was the creation of the High School for Teachers in 1880, with the aim of graduating teachers for high schools. Before this, the preparation of teachers concerned only the teaching of the Arabic language. With the advent of the High School for Teachers, the other academic subjects appeared among the curriculum, including mathematics. The High School was transformed into a High Institute for Teachers in 1929, later on to become a Faculty of Education at Ain Shams University in 1962.

Cairo University and Modern Egypt

Ahmed Lotfi El-Sayyed Pascha, 1872–1963

The Egyptian University started as a public, non-governmental institution near the end of 1908. Viewed by many as a symbol of rising Egypt, it was intended to rally the Egyptians around a huge cultural project. The demands for liberation and women's emancipation were growing stronger, alongside similar movements in many other countries. In this period of social anxiety and internal stress aroused by the occupation of Egypt by Great Britain, and also by international tensions preceding World War I, the renowned intellectual Ahmed Lotfi El-Sayyed emerged as a staunch advocator of liberalism and as a central figure of the Egyptian cultural and political life. El-Sayyed was appointed as the Vice-Chancellor of the Egyptian University from its very beginning, later on he became the Chancellor. The prevailing atmosphere was beneficial to the development of sciences, especially mathematics. In 1925, the Egyptian University became a state university. Alongside the faculties of art, law and medicine, the faculty of science was founded with several departments, among which were two mathematics departments.

A further step in the teaching of mathematics in Egypt was undertaken in 1926 when the Egyptian Government decided to hire one of the celebrated mathematicians of modern times, Professor Edward Lindsay Ince, with the task of building a strong department of pure mathematics at the Egyptian University (later to become King Fouad I University, then Cairo University since 1953). At that time, Ince was holding a post at the University of Liverpool. He stayed in Cairo for about five years before returning to England, during which the new department acquired its first curriculum. Differential and integral calculus, algebra, geometry, special functions, differential equations and the theory of functions of a complex variable constituted the basic subjects, in addition to others related to another department of applied mathematics: Newtonian mechanics, analytical mechanics and hydrodynamics (Ince biography, n.d.).

Ali Mostafa Mosharafa Pascha, 1898–1950

The first Egyptian to hold a Doctor of Science degree in mathematics was Ali Mostafa Mosharafa. Graduating from the Higher School for Teachers in 1917, Mosharafa was sent by the Egyptian government to England in 1917 to study mathematics. He was awarded the Bachelor of Science degree three years later, then the Doctor of Philosophy degree in mathematics in 1923 from King's College at the University of London. He got the Doctor of Science degree two years later from the same University.

Ali Mostafa Mosharafa was offered the post of lecturer at the Department of Applied Mathematics, Faculty of Science at the Egyptian University. Later on he became the first Egyptian Dean of this Faculty. He wrote several books on mechanics, theory of relativity, descriptive geometry, plane geometry and trigonometry, and many other books on different scientific topics for the large public. Mosharafa authored several papers on relativity and quantum mechanics, published in reputed scientific journals, e.g., *The Philosophical Magazine*, *Nature* and *Proceedings of the Royal Society of London.*

Mosharafa's contributions continued further. Reid (1990) states:

> Mosharafa knew that national scientific societies and journals are indispensable to a scientific community and did his best to foster them. When the Faculty of Science opened in 1925, the local scientific periodical press (aside from medicine and other applied sciences), consisted of little more than a bulletin of the Royal Geographical Society and the Institut d'Egypte—neither was dedicated mainly to natural science—and the occasional publications of the Entomological Society and the Helwan Observatory. Mosharafa encouraged the journal (founded in 1934) of his faculty, helped found the Mathematical and Physical Society of Egypt in 1936 (most of his later publications were in its proceedings), and backed the Egyptian Academy of Sciences (1944) and its Proceedings. Irregular and long delayed publications

> cut into the value of these and other scientific journals, and they circulated little outside of Egypt . . . Mosharafa reminded his audiences that Arabic had once been the international language of science for the Islamic and Mediterranean Worlds and he took seriously the task of reviving Arabic as a modern scientific language. He gave public lectures in Arabic, wrote Arabic mathematics texts and popular science books, and encouraged his colleagues to translate scientific textbooks into Arabic. In the Faculty of Science, Mathematics was the first department to have an all Egyptian staff and it led in introducing Arabic into the classroom. (p. 101–102)

Mosharafa wrote several books on geometry, trigonometry and mechanics for the final secondary stage at the Egyptian schools. He translated important textbooks in mathematics and mechanics, such as the *Differential and Integral Calculus* by Courant and Hilbert, and *Analytical Mechanics* by Burton.

Among the few Egyptians who got their Ph. D. Degree in mathematics during this period, one may cite Mohamed Ali Omara, who travelled to France on his own expenses and was granted the Ph. D. Degree in hydrodynamics from the University of Grenoble around 1928. Omara was offered a post of lecturer at the Department of Applied Mathematics, Faculty of Science at the Egyptian University, before becoming Head of Department in 1950–1951.

Mohamed Ali Omara among students at Assiut University in 1962.

Other Egyptian mathematicians who got their doctorate degrees in the late 1920's and early 1930's include Miltiady Hanna (Theory of Numbers, London) of Ain-Shams University, Amin Yasseen (Descriptive Geometry, London) of Cairo University, Ismail Adham (Quantum Mechanics, Moscow).

The first batch of Egyptian mathematicians to be graduated from the Faculty of Science at the Egyptian University was the class of 1929. Among the students of this class was Mohamed Morsi Ahmed, to later hold the post of Minister of Higher Education (1971–1972), and the General Secretary of the Union of Arab Universities in 1969–1971, then 1972–1980 ("أحمد محمد مرسي- ويكيبيديا، الموسوعة الحرة," n.d.).

Mohamed Morsi Ahmed

Mohamed Morsi Ahmed was granted the Ph. D. Degree from the University of Edinburgh in 1931. He became Professor of Pure Mathematics at the Egyptian University in 1943, Dean of the Faculty of Science in 1956, Vice-Chancellor of Cairo University in 1958, Chancellor of Ain-Shams University in 1961, and then Chancellor of Cairo University in 1967 ("أحمد مرسي محمد - ويكيبيديا، الموسوعة الحرة," n.d.).

Mohamed Morsi Ahmed translated many mathematical textbooks into Arabic. Besides, he was active in propagating scientific knowledge among the general public. He visited some countries in Europe and the United States under the auspices of the United Nations with the aim of transferring the European and American experiences in the field of statistics to Egypt in order to create an Institute of Statistical Studies in Egypt ("أحمد مرسي محمد - ويكيبيديا، الموسوعة الحرة," n.d.). Mohamed Morsi Ahmed helped in founding the University of Riyad in the Kingdom of Saudi Arabia.

The pure mathematics curriculum at the Cairo University (the heir of the Egyptian University) was modernized mainly due to the efforts of an outstanding Egyptian mathematician, Raouf Haleem Doss.

Raouf Haleem Doss
1915–1988

Doss took early studies in mathematics in Paris, France. He obtained a diploma in statistics in 1936, then a Bachelor Degree in mathematics in 1938. He got his M. Sc. Degree in 1942, then a Ph. D. in pure mathematics from King Fouad I University (later on Cairo University) in 1944. His thesis was supervised by the famous Hungarian-born mathematician Michael Fekete, who was also supervisor to the celebrated American mathematician John von Neumann. Raouf Doss was offered the position of assistant at the Department of Pure Mathematics, Faculty of Science, King Farouk I University (later on Alexandria University) in 1942, then a post of lecturer at the same department in 1944, before becoming associate professor in 1950.

THE QUARTERLY JOURNAL OF
MATHEMATICS
OXFORD SECOND SERIES
Volume 16 *No.* 62 *June* 1965

Raouf Doss: On de la Vallée-Poussin's proof of the fundamental theorem in the theory of almost-periodic functions 188

OXFORD
AT THE CLARENDON PRESS

A specimen of Doss publications

Raouf Doss taught mathematics at the Faculty of Higher Education at Baghdad, Kingdom of Iraq, in 1942. In the academic year 1945–1946, Doss attended a series of lectures by the renowned French mathematician Maurice Fréchét at Fouad I University. Doss solved one of the problems proposed by Fréchét to his audience and the result was published as "Sur la condition de régularité pour l'écart abstrait" in *Comptes Rendus* as early as 1946.

Doss visited the Institute for Advanced Study in New Jersey, USA, in 1949–1950 where he met John von Neuman. Raouf Doss moved to King Fouad I University in 1954 to hold the Chair of Statistics, then Chair of Pure

Doss receiving one of the national prizes in mathematics from King Farouk I in 1947.

Mathematics in 1958. During this period, Doss taught courses in mathematical logic, set theory, abstract algebra, general topology, measure theory and the theory of functions of a complex variable. He encouraged teaching modern courses on linear algebra, abstract algebra, differential geometry, harmonic analysis, functional analysis and others. Celebrated mathematicians visited the Department of Mathematics during Doss's period, including the Russian Andrey Kolmogorov, the Swiss Ernst Paul Specker, the Danish Børge Christian Jessen and the American Marshall E. Munroe.

In 1966, Raouf Doss accepted an offer for a professorship at New York University. He worked at Stony Brook campus for many years. Raouf Doss died Emeritus Professor in Setauket, New York, in 1988.

The curricula of applied mathematics at the Egyptian University were systematically reshuffled due to the acknowledged efforts of Ali Mosharafa and his student Mahmoud Bakri. New courses were added to the curricula, reflecting the interaction of the Egyptian faculty with the international scientific reforms. As early as 1959, Bakri taught the differential operators "grad," "div" and "curl" within a course of hydrostatics for the first year, potential theory for the second year, and introduced topics like optics and statistical mechanics for the third year at the Department of Mathematics, Faculty of Science, Cairo University.

Doss with the Indian mathematician Ram Srivastav on the occasion of Doss's retirement in 1986.

One of the central figures in Egyptian applied mathematics during the 1960–1990 period is undoubtedly Attia Abdel-Salam Ashour. His name is tightly connected to the subjects of electromagnetic induction in sheets, mixed boundary-value problems, dual series and dual integral equations, special functions and applications in geophysics.

Attia Abdel-Salam Ashour, 1924–

After his graduation from Fouad I University in 1944, Ashour was sent to Great Britain to further study mathematics. He got the Diploma of the Imperial College, London University in 1948, was granted the Ph. D. Degree from London University in 1948, and the D. Sc. from the same University in 1967.

During his stay in Great Britain, Ashour met and worked with celebrated scientists in the field of electromagnetism, like Sydney Chapman, Albert Price and Vincenzo C. A. Ferraro.

Ashour headed the Department of Mathematics at the Faculty of Science, Cairo University for long periods intermittently between 1959 and 1984. His extensive efforts in promoting the field of mathematics in Egypt and in Africa were highly esteemed. He was awarded many state prizes and Orders of Merit, Chevalier dans l'Ordre de la Palme Académique from the French Government in 1985, Medal of the African Union in 1990 and Chevalier dans l'Ordre National de Mérite from the French President in 1995. Ashour's international status was beneficial to the development of mathematics in Egypt and in Africa. He taught a course in boundary-value problems in Nigeria in 1969. His efforts led to the ratification of a scientific agreement between Cairo University and the Université Scientifique et Médicale de Grenoble for an extensive program of faculty exchange, under which the Department of Mathematics at Cairo University received several extended visits of renowned French mathematicians, while many Egyptian mathematicians from the department profited from scientific missions in Grenoble.

Ashour was member of many scientific local and international organizations. He acted as Vice-President of the International Union of

Geodesy and Geophysics from 1971 till 1975, then as President of this Union from 1975 to 1979. During the period from 1992 to 1996, Ashour was chosen President of the International Center of Pure and Applied Mathematics, Nice, France, and member of its Administrative Council from 1997 till 2000. In this capacity, he organized curricula in schools in Egypt. Ashour is presently Emeritus Professor at the Faculty of Science, Cairo University.

The Department of Mathematics at the Faculty of Science, Egyptian (Cairo) University helped teaching the mathematics curricula in practically all the newly founded universities around Egypt. In some cases, it totally supervised the new mathematics departments in these universities for many years, e.g., in Mansurah, Fayum and Beni-Soueif.

The Mathematical and Physical Society of Egypt

The Mathematical and Physical Society of Egypt was founded in 1936, due to the efforts of Ali Mostafa Mosharafa and Mohamed Morsi Ahmed of the Department of Mathematics, Faculty of Science at the Egyptian University, together with other renowned Egyptian mathematicians and physicists like Mohamed El-Nadi, Raafat Wassef and Helmi Youssef. The society soon acquired a scientific journal.

The National Committee for Mathematics

The National Committee for Mathematics was created within the Egyptian Academy for Science and Technology in the 1970's, mainly through the efforts of Professor Attia Abdel-Salam Ashour of the Department of Mathematics, Faculty of Science, Cairo University and others. This committee organized several national and international scientific conferences, as well as many workshops in different modern topics. Additionally, they convened to discuss the mathematics curricula in Egypt at grade school and university levels. The committee is composed of staff members mainly belonging to different faculties of science and engineering nationwide, so as to provide more insight into both the theoretical and applicable aspects of mathematics.

The Egyptian Mathematical Society

The Egyptian Mathematical Society was founded in 1992 through the efforts of Professor Afaf Sabri of the Department of Mathematics, Faculty for Women, Ain-Shams University and others, among whom Professor Abdel-Shafi Obada of the Department of Mathematics, Faculty of Science, Al-Azhar University, the actual President of the Society. The first volume of the scientific journal of the Society appeared in 1993. It is now part of Elsevier publishing system. The Society encompasses all the mathematics departments in Egypt in its activities and steering committee. It organizes one-day conferences and workshops, as well as a yearly international conference on different disciplines of mathematics.

Egyptian Mathematicians Outside Egypt

Scores of Egyptian mathematicians worked and are still working in universities and schools in practically all Arab countries. Their massive presence since the middle of the twentieth century was a key factor in the development of mathematics education in these countries. At university level, Egyptian mathematicians have made their valuable contribution as well, helping in the teaching of mathematics with Doss since 1945 in Iraq, Ashour in Nigeria and Bakri in Kuwait in 1970. These and other mathematicians helped to establish mathematics curricula in rising universities. Apart from individuals, Egypt had Cairo University Branch

Afaf Sabri among his students at Assiut University in 1960.

in Khartoum, the Sudan, from 1956–1993. In 1993, the University was "Sudanized" and renamed "El-Nilein University."

The Modern Egyptian Pre-University Educational System

In general, the Egyptian educational system has been subject to periodic revisions and development. Specialization of students starting from the second secondary school level was introduced in 1954–1955. In the period from 1955 till 1961, several topics were introduced into the mathematics curricula, e.g., imaginary quantities, the theory of the quadratic equation, statistics, descriptive geometry (drawing), solid geometry, and analytical treatment of the circle. Upon suggestion from UNESCO, a new curriculum for mathematics at the secondary school level was introduced in 1969 (Ebeid, 1980). The program of modern mathematics was directed by one the brilliant Egyptian mathematicians and statisticians, Abdel-Adheem Anis.

Abdel-Adheem Anis. 1923–2009

Anis graduated from King Fouad I University in 1944 and was granted the Ph. D. Degree from the Imperial College, London, in 1952. He was appointed Chair of Mathematics and Statistics at the Faculty of Science, Ain-Shams University in 1966. He presided the National Commission for the Development of Mathematics Education from 1970–1975.

A central figure in the modernization of the system of mathematics education in Egypt is undoubtedly Professor William Tadros Ebeid of the Faculty of Education, Ain-Shams University, whose continuous efforts are thankfully recognized by Egyptian mathematicians. Ebeid is considered to be the top of the pyramid in the field of mathematical education in Egypt. He obtained his Ph. D. Degree from the University

William Tadros Ebeid
1930–2012

of Michigan – Ann Arbor, USA. Ebeid played an active role in all reforms that took place in the educational system in Egypt. He frequently organized conferences and delivered lectures in Egypt and surrounding Arab countries, and he supervised theses on education in general and mathematics education in particular. His interest mainly focused on building of a national strategy for teaching mathematics to young students in light of well-defined modern standards.

In 2010, national standards were set for the Faculties of Education in Egypt concerning the teaching of mathematics and the training of mathematics teachers. These standards recognize the complexity of the world around us, and the role of the different branches of mathematics: mathematical analysis, algebra, geometry, probability and statistics, in dealing with it. They also point out the importance of logical thinking and proof, and to the branch of applied mathematics as an area of application of mathematics in dealing with and finding solutions for practical problems.

Mathematics Education Today

The Egyptian School Structure

The Egyptian educational system is currently made up of several stages: pre-primary education (2 years), primary education (6 years), preparatory education (3 years), secondary education (3 years) and post-secondary or University education. The primary and preparatory years have been compulsory for Egyptians since 1881, meaning that all students now receive mathematical training from the age of six until about the age of fifteen. The Ministry of Education oversees primary, preparatory and secondary education, while the Ministry of Higher Education oversees post-secondary education (Human Development Department Middle East and North Africa Region, 2007). Towards 1969, UNESCO proposed to Egypt the introduction of what is called "Modern Mathematics" into school curricula. This was implemented swiftly.

The level and type of mathematical education a contemporary student receives depends greatly on the type of school system they attend. School systems in Egypt are divided into governmental schools and private schools. There are two types of governmental schools: Arabic schools and language schools. Arabic schools teach the government curriculum in Arabic. Language schools teach most of the government curriculum in English, with only social studies being taught in Arabic (Human Development Department Middle East and North Africa Region, 2007).

The private schools can be categorized into four types. The first are private schools that follow a similar curriculum to government schools, and tend to differ only in the facilities they provide, such as newer buildings and teaching equipment. The second are language schools, which teach in English or in another language, usually French or German. At these schools mathematics is taught in the corresponding language. The third type is religious schools, such as Islamic or Catholic schools. These have some curricular independence, or run parallel with the government education system, such as Al-Azhar primary and secondary schools. Although, like the days of Muhammad Ali, these schools are deeply religious at their core, they usually implement the same mathematics curriculum that regular public schools use. Finally, there are international schools which follow another country's curriculum and system, such as British or French schools, or offer the International Diploma. The mathematics curriculum is these schools follows the county of origin's curriculum (Human Development Department Middle East and North Africa Region, 2007).

Mathematics Curricula

The next section highlights they key aspects of the current mathematics curriculum applied in all government schools in Egypt. It was obtained from the textbooks printed yearly by the Ministry of Education. Electronic versions of these textbooks can be found on many websites, one of which is http://kotbwzara.blogspot.com/.

Primary Education

Primary education in Egypt begins at the age of six with the introduction of numbers from 0 to 9. Students are taught the concepts of quantities and that numbers have order. The first year of school also includes addition and restricted subtraction of numbers 0 to 9.

The next few years involve similar concepts applied to larger numbers: three-digit and then five-digit numbers. The addition and subtraction of currency, measurement and units of length, and word problems are introduced in the second year of the primary stage. In addition, this level includes basic geometric concepts such as points and lines, shapes, edges, and number of lines. The third year builds on this by exploring 3-dimensional shapes, angles, and the use of rulers and protractors for length and angle measurement.

The six years of primary education end with students being able to manipulate fractions and common decimals, and understand ratios and proportions. Geometry at this level includes working with irregular volumes.

Mathematics Curricula for Preparatory Stage

The three years of preparatory mathematics education begin with the addition, subtraction, multiplication and division of fractions. The first year of this stage includes the introduction to algebra, including its application to operations such as calculating formulae for areas and unknown angles. The first year also includes basic concepts in statistics, such as means, medians and modes. The preparatory stage then progresses to topics such as the real numbers and Euclidean geometry.

Mathematics Curricula for the Secondary Stage

Algebra in the first year of the secondary stage includes algebra of matrices and roots of equations. Students also study trigonometric functions, including their inverses, properties and graphical representation. Finally, students study analytic and plane geometry.

Algebra during the second year includes exponents, logarithms, and their applications. Trigonometry includes sine and cosine rules for sum or difference of two angles, and trigonometric ratios of the double angle. This stage also introduces calculus with a focus on limits, variations of functions, rules of differentiation. Students are able to find the derivative of the product of two functions, the quotient of two functions, and trigonometric functions. Finally, the second secondary year sees a first introduction to mechanics: forces, equilibrium of a body under several forces, mean, instantaneous and relative velocity, accelerated motion, and derivatives of vector functions.

The third year of the secondary phase includes the study of permutations and combinations and the study of complex numbers (including roots and exponential forms) within the algebra section. It also includes further exploration of differentiation and integration: local and absolute maxima and minima, points of inflection, properties of integrals, integrals of trigonometric functions and their applications. Students also cover a range of topics within the study of statics and dynamics, including motion on a rough plane, work, power, kinetics and potential energy. Finally, students study a number of concepts in statistics including probability distributions, correlations and regression.

References

Academy of Scientific Research and Technology. (1993). *Tareekh al-uloom wal-technologia al-handasiya fi masr fil-qarnayn al-tase' ashar wal ashrayn* [History of science and engineering technology in Egypt in the nineteenth and twentieth century] (1).

Al-Azhar University. (n.d.). Retrieved from http://www.azhar.edu.eg/En/u.htm

Ashour, A. A., Khalil, A. B. & Hassan, N. A. (1990). *Tareekh al-haraka al-ilmiya fi masr al-hadeetha* [The history of the scientific movement in modern Egypt] (2). Academy of scientific research and technology.

Ebeid, W. (1980). *Kalima fi a'mal wa tawsiyat mo'tamar ta'lim al-riyadiyat li-marhalat ma qabl al-jami'a* [A word on the work and recommendations of the conference on mathematics education for the pre-university stage]. Academy of scientific research and technology.

Eves, H. (1990). *An introduction to the history of mathematics* (6th ed.). CA: Brooks/Cole.

Human Development Department Middle East and North Africa Region. (2007). Arab Republic of Egypt Improving Quality, Equality, and Efficiency in the Education Sector (42863-EG). Retrieved from World Bank website: http://www-wds.worldbank.org/external/default/WDSContentServer/WDSP/IB/2008/06/26/000334955_2008062603 3032/Rendered/PDF/428630ESW0P08910gray0cover01PUBLIC1.pdf.

Ince biography. (n.d.). Retrieved from http://www-history.mcs.st-and.ac.uk/Biographies/Ince.html.

Joseph, Baron Fourier | French mathematician | Britannica.com. (n.d.). Retrieved from http://www.britannica.com/biography/Joseph-Baron-Fourier.

Reid, D.M. (1990). *Cairo University and the making of modern Egypt*. Cambridge University Press.

Tawqan, Qadri H. (2008). Turath al-Arab al-elmi fi al-riyadhiyat wal-falak [Arab scientific heritage in mathematics and astronomy]. Cairo: General Organization of Cultural Palaces.

About the Authors

Ahmed Fouad Ghaleb obtained his Ph.D. in Mathematics and Physics from Moscow State University in 1976 in the field of Continuum Mechanics of Electromagnetic Media. He is currently Professor Emeritus at the Department of Mathematics, Faculty of Science, Cairo University. His actual interests include the History of Mathematics in Egypt and the Translation into Arabic of the Mathematical Terminology.

Soha Abdelsattar is a Ph.D. student in the mathematics education program at Teachers College, Columbia University. She was born and raised in Doha, Qatar, and in 2010 she earned her bachelor's degree in mechanical engineering from Texas A&M University at Qatar. Through witnessing a large scale education movement in Qatar, she was inspired to enter the field of education herself, and in May 2013 she obtained a master's degree in Bilingual and Bicultural Education from Teachers College. She then enrolled in the Ph.D. program in mathematics education at Teachers College. As a Qatari citizen of Egyptian heritage, her interests include mathematics and its teaching in the Middle East and North Africa (MENA) region, specifically with regards to the history of mathematics and its role in shaping Middle-Eastern curricula today.

https://doi.org/10.1142/9789813146785_0004

Chapter IV

INDONESIA: The Development of Mathematics Teacher Education in Indonesia

Abadi
Mathematics Department, Universitas Negeri Surabaya (Unesa), Indonesia

Zahra Chairani
Sekolah Tinggi Keguruan dan Ilmu Pendidikan (STKIP)
Persatuan Guru Republik, Indonesia (PGRI)
Banjarmasin, Kalimantan Selatan, Indonesia

Introduction

The history of education in Indonesia developed alongside the nation itself. Indonesia was called Nusantara when the monarchy era was established. Education in that era was brought by merchants from India and China who also acted as preachers introducing Hinduism and Buddhism to the society. In that era, many kingdoms developed with Hindu or Buddha as their background, such as Tarumanegara, Kutai, Sriwijaya, and Majapahit. Preachers and monks introduced Hinduism and Buddhism by teaching people about godhead, humanity, and science as well. As a result, those kingdoms bequeathed many historical artifacts such as temples and other architectures and cultures. Those could be considered the products of education at that time that contributed to socio-cultural life in present day Indonesia.

Following that era around the tenth century, when Islamism was brought to Indonesia by Arabic merchants, another system of education was introduced to society. Islamic preachers spread Islamic religious teachings through mosque or *pesantren* community where people could discuss Islam and other knowledge. That process of teaching continuously developed in Islamic kingdoms such as Pasai in Aceh and Bandar in Kalimantan (Hasbullah, 1999; Rukiati, 2006). So, in that era, Indonesia consisted of societies with different beliefs and faith, i.e. either Hinduism, Buddhism, Islamism, or other faiths. Amazingly, the divergent societies could live together in harmony and respect with one another.

Living together with many differences was too risky and endangered the societal harmony. The whole society gradually became unstable. Friction between various groups happened when there was an external factor that influenced the harmony in the society. For example, during the colonialism era in 17th century, when Portuguese, Spanish, and Dutch merchants entered the territory, they brought missions, Christianity, their cultures, and expanded trading networks. Due to a long period of occupation by the Dutch (about 350 years) Indonesia inherited many things brought by the Dutch, including their education system.

Having such long and varied experiences of different influences, after its independence, Indonesia became a nation that was united in the diversities it had, including ethnicities, religions, languages, and cultures. Those factors predominantly contributed to the development of the country, including the development of education. It can be seen that the education system in Indonesia is a unique one. Besides the secular education system, inherited from the Dutch, where most students in the country attend, there is also an Islamic education system to accommodate some of the Muslim students.

The Establishment of Teacher Education

The development of education cannot be separated from the development of teacher education. Likewise, the development of teacher education in Indonesia cannot be separated from the long history of the country. The demand for teachers with good qualifications resulted in some changes of the teacher education system from period to period.

The education system for teachers was informally established in Hinduism, Buddhism, and Islamic eras. The system was rooted in religious communities where the teaching and doctrines of the religions were preached. In that stage, some pedagogical activities were also practiced, and those practices become hereditary through generations. Particularly in the Islamic era, the education system was already established, although in a traditional way. School levels, such as elementary school, secondary school, and higher education, were available. This was developed in Islamic kingdoms, such as Pasai in Aceh, Bandar in Kalimantan, and Kembar Gowa Tallo in Sulawesi (Hasbullah, 1999). The curriculum mostly consisted of Islamism and Qur'an studies. The system established by Islamic schools had already provided a good partition of levels, including higher education.

The coming of the Dutch to colonize the country hampered the development of the Islamic education system in Indonesia. The Dutch believed their mission was not only for trade expansion, but also the spread of Christianity. That was the reason they did not allow Islamic education system developed in the country. Instead, the Dutch introduced a formal education system adopted from their own country. However, the implementation of the system was not entirely the same as the original one, since the Dutch discriminated between the native Indonesian and the Dutch. This was because there were political agendas behind the decision to introduce education to Indonesia, so that only a certain class (aristocrats) of native origin could enjoy the schooling. Nevertheless, from those experiences Indonesians started to learn about formal education systems. They learned about levels of education from elementary through higher education. They also learned the importance of higher education, because from that level of education they could prepare professionals such as engineers, teachers and medical doctors.

In the Dutch colonialism era, Indonesia learned how to prepare teachers. Since schools were stratified into divisions, schools for native pupil, schools for Chinese, and schools for the Dutch and elite natives, teacher education was differentiated depending on where the prospective teachers would teach. Education for teachers who would teach Folk School (three-year elementary school for native pupils) was conducted in two-year *Cursus voor Volk Onderwijs* (CVO) and education for those who would teach secondary elementary school (five-year elementary school for native pupils) was conducted in four-year Normal Schools. The candidates of those schools had to graduate from elementary school. Meanwhile, for those who would teach in *Holland Inlandsche School* (HIS), a seven-year elementary school for the Dutch taught in Dutch language) must take HIK for another six years after HIS. Those who taught secondary school (*Meer Uitgebreid Lager Onderwijs* (MULO)), took *Hoofdt Acte* (Kelabora, 2006).

Despite continuous disapproval from the Dutch, Islamic teacher education also developed (albeit in a traditional way). Islamic scholars called *Wali Songo* played a central role to development of Islam, particularly in Java Island (Fathurrohman, 2012). They introduced Islamism through interesting teaching methods that assimilated Islamic teachings into Javanese cultures. These Islamic teaching methods were believed to be at the root of the pedagogy used in Islamic education institutions like *pesantren*.

Japanese occupation of Indonesia during World War II provided not only three and a half years of misery to Indonesia; it also shaped the education system. Japanese policy regarding education in Indonesia "helped" to abolish the dualism in education introduced by the Dutch. They required the use of Indonesian language, rather than Dutch, in schools. These might be factors that encouraged Indonesian pupils to be independent by the end of the World War II, August 17, 1945.

Mathematics Education and Mathematics Teacher Education

The development of mathematics teacher education in Indonesia cannot be separated from the curriculum of school mathematics used in Indonesia since independence. The Indonesian government continued the existing education system inherited from both the Islamic education system and the secular education system brought by the Dutch, with some modifications. Indonesia adopted those two education systems, each under the supervision of two different ministries. The Islamic education system is supervised by the Ministry of Religious Affairs (MORA). Meanwhile, the secular education system is supervised by the Ministry of Education and

Cultures (MOEC, formerly MONE). Although supervised by different ministries, both education systems are inseparable in the context of the national education system because they follow the same curriculum that applies to all schools.

The country had already experienced at least seven curricula since 1968 (Mailizar, 2014). Each curriculum gave different emphasis on either content or approaches depending on the competencies being achieved. Earlier curricula put more emphasis on mathematics content, while later curricula emphasized the compatibility of mathematics content with the cognitive level of students and teaching methods/approaches being used.

In Curriculum 1968, mathematics was based on the learning theory of B. F. Skinner (Ruseffendi, 1988). To strengthen students' understanding about certain topics, exercises for affirmation after stimulus-response processes were carried out. Thus, the teaching and learning process was more teacher-centered and rote and arithmetic-heavy, but less motivating. A large change in the approach to teaching mathematics occurred when modern mathematics was included in Curriculum 1975. It was due to the use of mixed learning theory of associations forming between stimuli and responses by R. L. Thorndike combined with the theory of cognitive development by I. Piaget and behaviorism by B. F. Skinner and R. Gagne (Ruseffendi, 1988). The teaching and learning process was student-centered, giving more priority to understanding and problem solving skills, and not rote-based anymore. Curriculum 1984 did not change much from the previous curriculum except for introducing computers and calculators into mathematics curricula. In Curriculum 1994, another restructuring was done, especially of some mathematics contents, such as special emphasis on number sense in elementary school, and introducing graph theory in senior high school. Rigorous deductive thinking was also introduced, particularly in geometry. Curriculum 2004 was also called "competence-based curriculum." It refined Curriculum 1994 especially by applying learning theories of constructivism and cognitive development. Moreover, Curriculum 2004 put emphasis on developing problem solving ability, logical, critical, and creative thinking, and, in addition, on communicating mathematical ideas. Regarding all aspects implemented in Curriculum 2004, Curriculum 2006, which was also called school-based curriculum (BNSP, 2006), gave more freedom to teachers to develop their school's own curriculum in accordance with the national standard of competencies provided by the government. The curriculum adapted the four pillars of education recommended by UNESCO (Singh, 2011) that also required attitudes and values as parts of the students' competencies. Finally, Curriculum 2013, despite its controversy, was more systemic, flexible, and contextual in character. It provided core competencies and base competencies for each subject. Core competencies were categorical descriptions of competencies of students at a certain level of school, class, and subject. Meanwhile, base competencies were competencies that had to be learned by students in a subject (Depdikbud, 2013; Mulyasa, 2013; 2014). To implement, the curriculum teachers were told to use a scientific approach, which was not always easy to implement. The scientific approach consists of activities such as observing, asking, investigating, associating, and communicating.

With all of these changes in the school mathematics curriculum, appropriate changes were also made to the curriculum of mathematics teacher education. Due to the fact that Indonesia implemented dualistic systems, the government also provided two teacher education institutions. Each of these institutions is under those two ministries supervision.

The aim of developing teacher education was to meet the need for qualified teachers for schools of all levels. Like school curricula, teacher education transformed from time to time. For example, teacher education for elementary schools was transformed from Normal School into Teacher School B, which was four years long as well. In 1957 another transformation occurred, turning the school into a three-year Senior Teacher School (*Sekolah Guru Atas*). This eventually became the School of Teacher Education (SPG) that required candidates to be graduated from junior high school instead of from elementary school. Later in the 1980s, elementary school teacher candidates were required to be graduated from senior high school and take a two-year diploma program provided by the Institute of Teacher Training (LPTK). Similar situations also happened for teacher education for secondary school. Eventually, teacher education for secondary school was merged into the Institute of Teacher Training (LPTK). In the 1990s, all teacher education was combined with the undergraduate (S1) program under the management of the LPTK (Depdiknas, 2013).

The process of transformation of the Islamic education system was not quite the same as that of the secular education system. Due to strong disapproval from the Dutch in the colonialism era, the transformation of the Islamic education system moved rapidly after independence. It was the Ministry of Religious Affairs that was responsible for transformation of the system. The Ministry not only prepared teacher education for general subjects but also teacher education for Islam as a religion (Kosim, 2007). The stages of transformation of teacher education in Islamic education system generally followed the model of the secular system.

Mathematics teacher education in Indonesia was transformed by the above process. First, the curriculum of mathematics teacher education was aligned with that of school mathematics with some additional higher mathematical topics and pedagogy depending on the competencies and profile of the graduates. Mathematics teacher education under the administration of Institute of Teacher Training (LPTK), originated from programs B1 and B2, focused on mathematics, sciences, English, German, engineering, economics, and physical education, topics inherited from the Dutch (Kelabora, 2006).

Professional Development for Teachers

In 1980, due to fast developments in technology, calculators and computers were integrated in mathematics education. To accommodate this change, the government introduced in- and on-service training model for mathematics teachers, with the objectives of the training as follows:

1. Upgrading professional mathematics teaching.
2. Introducing the use of technology, such as calculators and computers, in certain mathematics subjects.
3. Understanding the connection between mathematics teaching in elementary school, secondary school and higher education.
4. Improving mathematics teaching for active learning.
5. Evaluating the teaching and learning process in mathematics education.

The training was aimed to improve the quality and performance of mathematics teachers. The project started with 30 teachers who were selected from 10 provinces. The selected teachers were intended to become instructors of other teachers. They were trained in several countries, such as Malaysia, Scotland, and England. Trainees attended some short courses and activities related to designing mathematics lesson plan, choosing appropriate assessment and evaluation methods, and designing remedial programs. Upon return to their home province, each instructor was responsible for disseminating the knowledge she or he obtained from the training. For one academic year, each province had to train 40 mathematics teachers. The training was two semesters long; in the first semester the teacher trainees were trained for two weeks (14 days) followed by implementation in each school for 12 weeks, then group evaluation and individual evaluation for one week, and, finally, another two-week implementation in schools. The second semester followed the same procedure as the first. Training materials mainly focused on learning a new approach to teaching mathematics and changing from teacher-centered to active learning paradigms. The model was as illustrated in Figure 1.

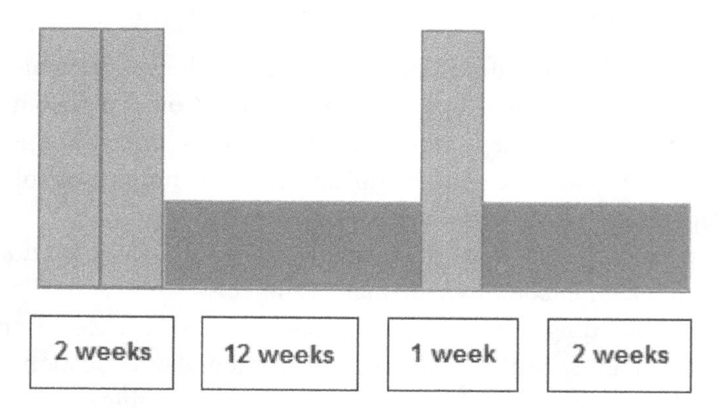

Figure 1. In-service and on-service model of teacher training model for one semester of 17 weeks.

Due to a lack of facilities and resources, the project only survived until 1995. Many revisions had been made to the pattern of teacher training and the recruitment systems, which was directly under the control and supervision of the Ministry of Education and Culture, to accommodate curriculum changes in 2004, 2006 and, finally, in 2013.

Mathematics teachers' professional development is also carried out through non-structural teachers' organization called *Musyawarah Guru Mata Pelajaran* (*MGMP*), mathematics teachers' association and *Kelompok Kerja Guru* (*KKG*), or mathematics teachers' workshop group. The association and group are forums for mathematics teachers to discuss and share problems encountered in teaching and learning mathematics, including lesson plan preparation, teaching materials, and issues on the implementation of a new curriculum. Due to its central role for supporting teachers' professionalism and the implementation of Decree Number 14 Year 2005 about Teachers and Lectures, the Department of National Education provided the standard of development for *MGMP* and *KKG* (Depdiknas, 2008) to guide teachers in forming their own *MGMP* or *KKG.*

As the need of professional teachers is a necessity, recognition of teaching as a profession is required. Therefore, there are some competencies requirements for teachers so that they can be accepted by the environment academically, pedagogically, and professionally while they are in service. To be professional, teachers need to be prepared through a program that is designed based on teachers' competencies. For that purpose, the Indonesian government formed an institution to undertake education of teachers' profession (Ari Widodo and Riandi, 2013).

There are two decrees which establish the juridical base for the institution of teachers' profession, namely, decree No. 20 Year 2003 about the System of National Education (Depdiknas, 2003), which constitutes that education of teachers' profession is higher education after an undergraduate (S1) program that prepares students to have an occupation requiring special skills. Decree No. 14 Year 2005 about teachers and lectures (Depdiknas, 2005) constitutes that teachers are professionals in all levels of education (Depdiknas, 2005). In addition, the government designated that the program is to be conducted by the Institute of Teacher Training (LPTK). The program is mainly focused on strengthening teachers' competencies as constituted in decree No. 19 Year 2005 about the National Standard of Education (Depdiknas, 2005), saying that teachers' competencies are pedagogical competencies, personality competencies, social competencies, and professional competencies. Therefore, professional teachers' education is aiming to prepare graduates that:

- are able to show a set of competencies which are in line with the standard applied.
- are able to apply scientific and technological tenets where they are in-service
- comply with ethical codes of teacher profession.
- are able to work with full dedication.
- are able to make decisions independently and collectively.
- are able to show accountable performance.
- are able to work together with colleagues and other relevant parties.
- continuously make self-improvement independently or through profession association.

In order to achieve those goals, a set of curricula for teachers' professional education is needed. The curriculum being constructed must consider the input of previous knowledge. Since the input of the program can be both those who have education background (LPTK graduates) and those who do not (non-LPTK graduates), the curriculum of the program consists of two tracks, for LPTK graduates and for non-LPTK graduates. Table 1 shows curriculum structures of the program for each of the groups.

This curriculum is planned for two semesters; the first semester covers academic competencies and the second semester covers professional competencies.

Table 1. Curriculum structures of the program of Education of Teachers' Profession (Depdiknas, 2008).

No	Competencies	LPTK graduates	Non-LPTK graduates
1	Academic	Strengthening and preparing subject matter for teaching and learning (subject specific pedagogy)	Study on pedagogical knowledge Study on students' development Preparing subject matter for teaching and learning (subject specific pedagogy) Preparing teachers' personality competencies
2	Professional	Teaching internship	Teaching internship

Mathematics Education at The State University of Surabaya

In the State University of Surabaya (Unesa), the Mathematics Education Study Program is part of the Department of Mathematics, of the State University of Surabaya. Surabaya is the second biggest city in Indonesia, about 1000 kilometers to the east of Jakarta, the capital city of the country. The city is located in the East Java province and situated with a harbor at the East and with mountains at the south.

Historically, Unesa originated from LPTK and called IKIP. In 1999, Unesa got a wider mandate to be the State University of Surabaya. As a university, Unesa is authorized to run not only an educational study program but also a non-educational study program. Likewise, The Department of Mathematics also has two study programs, the S1 (undergraduate) Mathematics Study Program and the S1 Mathematics Education Study Program. It is mandated by decree No. 8 Year 2012 about Indonesian Qualification Framework (IQF) (Depdiknas, 2012) and decree No. 45 Year 2014 about National Standard of Higher Education (NSHE) (Depdiknas, 2014). The curriculum of every study program has to be based on the Indonesian Qualification Framework and the National Standard of Higher Education. So do the curricula of the two study programs.

The framework is needed especially when graduates of a study program enter the global market, seeking higher education or jobs overseas, where education classification is required. The framework does not only apply to formal education but also accommodates experience-based learners to be granted a degree through a certification program. Meanwhile, the National Standard of Higher Education provides learning outcomes related to the attitude, general skills, knowledge, and specific skills that must be achieved by graduates of the study program. The standard already provides the learning outcomes on attitude and general skills. Study programs may add more learning outcomes to provided outcomes, depending on the local values that the study program has. Learning outcomes on knowledge and specific skills are each study program's responsibility and are communicated with study programs nationwide.

The Mathematics Education study program of Unesa has prepared its curriculum following the guidelines provided by IQF and NSHE as mentioned above. The one-year process of curriculum development included the following stages: tracer study, comparative study, curriculum evaluation, designing a new curriculum, sanctioning and public hearings and trials. During the stages before designing a new curriculum, the study program communicated with other mathematics education study programs nationwide under the association so called Indonesian Mathematics Society (IndoMS). The forum primarily discussed the minimum curriculum that all mathematics education study programs in the country must have. At that stage, the association formulated the minimum learning outcomes on knowledge and specific skills equipped with the list of courses. Before designing its curriculum, the study program decided the profiles of the graduates based on the potential of the study program. According to the results of tracer study, comparative study, and self-evaluation, the study program decided that the profiles of the program graduates are:

1. Teachers of secondary school
2. Mathematics education researchers
3. Edupreneurs

Based on the profile of these graduates being, the Mathematics Education study program of Unesa formulated its learning outcomes as guided by IQF and NSHE as shown in Table 2.

Table 2. Learning Outcomes of S1 Mathematics Education Study Program.

PARAMETER	STUDY PROGRAM LEARNING OUTCOME
Attitude	Cautious to God almighty and be able to show religious manner
	Upholding humanity values while on duties by virtue of his religion, moral and ethics
	Contribute in improving the society's quality of life and the advancement of civilization by virtue of Pancasila; the five principles of the nation
	Play role as a citizen who proudly loves being Indonesian, having nationalism and is responsible to the country and the nation
	Value the diversity of cultures, views, religions, beliefs, and opinions or genuine findings of others
	Cooperative, having social sensibility, and environmental concern
	Law-abiding and discipline in social life and good citizenship
	Internalize academic values, norms, and ethics
	Show responsible character on the work of his/her specialty independently
	Internalize spirit of self-reliance, hard work, and entrepreneurship
	Actualizing the characters of faith, intelligent, self-reliant, honest, care, and tough in daily conduct
	Having sincerity, commitment, and wholeheartedness to develop students' attitude, values, and competencies
General Skills	Be able to apply logical, critical, and systematic thinking in the context of developing or implementing science and technology that considers and applies humanity values in her/his area of specialty
	Showing genuine, quality, and measurable pieces of work
	Be able to examine the implication of development and implementation of science and technology that considers and applies humanity values based on rules, methods, and ethics in order to find solutions, ideas, designs, or art critiques.
	Be able to provide a scientific description of results of study in the form of theses or final reports, and upload them on the university website.

PARAMETER	STUDY PROGRAM LEARNING OUTCOME
	Be able to make correct decisions in the context of problem solving in the area of her/his specialty based on information and data analysis
	Be able to maintain and broaden network with supervisor, internally and externally network
	Responsible for group achievement and be able to supervise and to evaluate on her/his staff performances
	Be able to do a self-evaluation process toward groups under her/his supervision and manage teaching and learning processes independently
	Be able to document, save, secure, and reload data for the purpose of validity and prevention from plagiarism
Knowledge	Comprehend pedagogical-didactical concepts for teaching school mathematics that are oriented to life skills
	Comprehend theoretical concepts of mathematics including: logics, discrete mathematics, algebra, analysis, geometry, probability and statistics, mathematical modeling principles, linear programming, differential equations, and numerical methods that support mathematics teaching in elementary and secondary schools, and further study as well
	Comprehend principles and technique of planning, managing, and evaluating mathematics teaching and learning
	Comprehend factual knowledge of functions and benefits of using technology especially relevant information and communication technology for teaching mathematics
	Comprehend research methodology for conducting research on mathematics education
Specific Skills	Be able to plan, implement, and evaluate mathematics teaching and learning innovatively by applying pedagogical-didactical concepts and mathematics as subject matter, also by utilizing various resources that are oriented to life skills
	Be able to examine and apply various teaching methods innovatively and reliably
	Be able to conduct tutorials for students
	Be able to plan and conduct research to find alternative solutions of problems in the topic of mathematics education and publish the results

NOTE: *italic face means that the learning outcomes are the hallmark of Unesa.

The formulation of the learning outcomes on knowledge and specific skills resulted from intensive discussion among lecturers of the study program and communicated the results in the forum of mathematics education study programs. From those learning outcomes of the study program, courses' learning outcomes were derived. The courses' learning outcomes were used as the base to formulate courses that are needed by the intended profiles. Table 3 lists the courses offered by Mathematics Education study program of Unesa compared to the minimum curriculum recommended by the IndoMS.

Table 3. Corresponding courses offered by Mathematics Education study program of Unesa compared to the minimum curriculum recommended by IndoMS

No.	Field of Study	IndoMS		Mathematics Education study program of Unesa	
		Sub field/ Courses	**Credit**	**Courses**	**Credit**
1.	Matematika	Basic Mathematics	3	Basic Mathematics	3
		Analysis	6	Real Analysis I	3
				Multivariable Calculus	3
		Geometry	6	Geometry	3
				Analytical Geometry	3
		Algebra	6	Abstract Algebra I	3
				Elementary Linear Algebra	3
		Statistics	6	Statistical Method	3
				Probability and Statistics	3
		Applied Mathematics	6	Operation Research	3
				Mathematical modeling	3
		Calculus	6	Differential Calculus	4
				Integral Calculus	4

No.	Field of Study	IndoMS		Mathematics Education study program of Unesa	
		Sub field/ Courses	Credit	Courses	Credit
		Number Theory	2	Elementary Number Theory	2
		Differential Equations	3	Ordinary Differential Equations	3
		Discrete Mathematics	3	Discrete Mathematics	3
		Numerical Analysis	3	Numerical Method	3
2.	School Mathematics	Selected topics in Elementary school mathematics	3	The study of elementary school curriculum	2
		Selected topics in Secondary School Mathematics	3	The study of secondary school curriculum	3
3.	Mathematics Learning	The planning of mathematics learning	3	Development of teaching materials	3
		Methodology of Mathematics Teaching	3	Innovative teaching and learning I	3
				Innovative teaching and learning II	3
		Evaluation of Mathematics Teaching and Learning	3	Assessment of Mathematics Teaching and Learning	3
		Teaching Media	3	Teaching Media	2
				Computer Application for teaching and learning	2
		Research Methodology	3	Educational Research Methodology	3

No.	Field of Study	IndoMS		Mathematics Education study program of Unesa	
		Sub field/ Courses	**Credit**	**Courses**	**Credit**
4.	Pedagogy	The Fundament of Education	2	The Fundament of Education	3
		Psychology of Education	2	Psychology of Education	2
		Curriculum and its learning	2	Learning Theory	3
		Education Management	2	Teaching Internship	3
		Teaching internship	2		
5.	Thesis	Thesis	6	Thesis	6
		Total	**87**		**93**

As can be seen in Table 3, there are some differences between the names and credits recommended by IndoMS with those offered by the study program. It is possible that, because the study program has already fulfilled the minimum requirement asked by the association, they then deepened some content in the courses. By doing so, the courses' names were changed and the credits were added as well. The study program needs to add additional courses to meet the needs of the profiles being set. A student is approved to graduate from the study program if she or he has already taken at least 144 credits with a certain composition of courses.

Conclusion

Due to its historical background, Indonesia became a country that has a dualistic education system. With all of its diverse characteristics, the Indonesian government managed to develop a system of education that facilitates the need of good education for all. Interestingly, although most schools of all levels follow a secular education system, the system still accommodates learning outcomes on attitudes that were adopted from the Islamic education system, particularly about divinity, faith, and beliefs.

In the globalization era, the Indonesian government is confident that the education system being implemented can produce graduates that are ready to compete in global market. The policy of implementing the Indonesian Qualification Framework (IQF) equipped with the National Standard of Higher Education (NSHE) is a smart and well-planned strategy to face the globalization era. By using the framework and standards, higher education institutions in the country have an opportunity to formulate their curriculum to meet global market standards and promote local values that support national resilience at the same time.

In implementation, the framework and the standards apply well to any study program of higher education. The mathematics education study program of Unesa which has applied the Indonesian Qualifications Framework (IQF) and NSHE in developing its new curriculum. The new curriculum is designed to reflect the study program's strength and potential to produce prospective mathematics teachers to meet the global market.

Acknowledgements

Many thanks to the Mathematics Department of Unesa that gave the corresponding author great access to the curriculum of the Mathematics Education study program.

References

Ari Widodo & Riandi, 2013. Dual-mode teacher professional development: challenge and revisioning future TPD in Indonesia. *Teacher Development: An International Journal of Teachers' Professional Development*, Volume 17 (3), pp. 380–392.

BNSP. 2006. *Panduan Penyusunan Kurikulum Tingkat Satuan Pendidikan: Jenjang Pendidikan Dasar dan Menengah* (in Indonesian). Jakarta.

Fathurrohman, M. 2012. *Sejarah Pendidikan Islam di Indonesia.* (in Indonesian). https://muhfathurrohman.wordpress.com/2012/09/15/sejarah-pendidikan-islam-di-indonesia (accessed on December 1, 2015).

Depdikbud. 2013. *Peraturan Menteri Pendidikan dan Kebudayaan Nomor 81A tentang Implementasi Kurikulum.* (in Indonesian).

Depdiknas. 2008. *Standard Pengembangan Kelompok Kerja Guru (KKG), Musyawarah Guru Mata Pelajaran (MGMP).* (in Indonesian). Jakarta.

Depdiknas. 2008. *Pendidikan Profesi Guru Prajabatan* (in Indonesian). Jakarta.

Depdiknas. 2013. *Menyiapkan Guru Masa Depan* (in Indonesian). Jakarta.

Depdiknas. 2003. Undang-Undang Republik Indonesia Nomor 20 Tahun 2003 tentang Sistem Pendidikan Nasional (in Indonesian). Jakarta.

Depdiknas. Peraturan Pemerintah Republik Indonesia Nomor 19 Tahun 2005 tentang Standar Nasional Pendidikan (in Indonesian). Jakarta.

Depdiknas. Peraturan Presiden Republik Indonesia Nomor 8 Tahun 2012 tentang Kerangka Kualifikasi Nasional Indonesia (in Indonesian). Jakarta.

Depdiknas. Peraturan Menteri Pendidikan dan Kebudayaan Republik Indonesia Nomor 49 Tahun 2014 tentang Standar Nasional Pendidikan Tinggi (in Indonesian). Jakarta.

Hasbullah. 1999. *Sejarah Pendidikan Islam di Indonesia; Lintasan Sejarah Petumbuhan dan Perkembangan* (in Indonesian). Raja Grafindo Persada. Jakarta.

Kosim, M. 2007. Dari SGHAI ke PGA: Sejarah Perkembangan Lembaga Pendidikan Guru Agama Islam Negeri Jenjang Menengah (in Indonesian), *Tadris*, Volume 2. Nomor 2. pp. 279–196.

Kelabora, L. 2006. The Evolution of Teacher Education in Indonesia. *South Pacific Journal of Teacher Education*. Volume 3 (1), pp. 33–43.

Mailizar. 2014. A historical overview of Mathematics curriculum reform and development in Modern Indonesia. *Teaching Innovations*. Volume 27 (3). pp. 58–68.

Mulyasa, E. 2013. *Kurikulum Berbasis Kompetensi, Konsep, Karakteristik dan Implementasi*. (in Indonesian). Bandung: Remaja Rosdakarya, pp. 8–42.

Mulyasa, E. 2014. *Pengembangan dan Implemenasi Kurikulum 2013* (in Indonesian). Bandung: Remaja Rosdakarya. pp. 20–45.

Rukiati, E.K. and Hikmawati, F. 2006. *Sejarah Pendidikan Islam di Indonesia*. Pustaka Setia. Bandung.

Ruseffendy, E. T. 1988. *Pengantar Kepada Membantu Guru Mengembangkan Kompetensinya dalam Pengajaran Matematika untuk Meningkatkan CBSA* (in Indonesian), Bandung.

Singh, J. P. 2011. *United Nation Educational, Scientific, and Cultural Organization (UNESCO) Creating norms for a complex world*. New York.

About the Authors

Dr. Abadi is currently Head of Department of Mathematics Departement of the State University of Surabaya (or in Indonesian, Universitas Negeri Surabaya (Unesa)). He received his B.Sc. from IKIP Surabaya (the former name of Unesa), Indonesia, his M.Sc. in mathematics from The University of Queensland, Australia, and his Doctorate in applied mathematics from Universiteit Utrecht, the

Netherlands. His research interests are in dynamical systems as well as in mathematics education as he produced a number of papers in both areas. Dr. Abadi is a member of the Indonesian Qualification Framework of mathematics division and he is one of the chairman of the Indonesian Mathematics Society (IndoMS) of East Java Province. Dr. Abadi loves to play tennis, cycling, and watch football matches. He lives in Surabaya with his wife and their two sons.

Dr. Zahra Chairani, M.Pd. is a lecturer in Mathematics Education Program of STKIP PGRI Banjarmasin, South Borneo, Indonesia. She received her bachelor's degree of mathematics education from Universitas Negeri Lambung Mangkurat (UNLAM) Banjarmasin in 1981, and her M.Pd. – Magister Pendidikan (Master of Education) Universitas Negeri Malang (UM) Malang, East Java, in 2002, and her Ph.D. in Mathematics Education from Universitas Negeri Surabaya (UNESA), East Java, in 2015. Her dissertation focused on Students Metacognition in problem solving. She attended short programs of mathematics teacher training in SEAMEO RECSAM, Penang, Malaysia (1982), Dundee College, Scotland (1983) and King College, London, England (1997).

Zahra taught mathematics at a senior high school in Banjarmasin (1969–1993). She was one of the qualified trainers for the mathematics teachers in Indonesia for Banjarmasin (1982–1995). She later joined the Insurance Institute for Qualify Education (LPMP) in 1993 and became a consultant of Mathematics Innovation in 2005–2007 for the Ministry of Education. She received a reward in 2007 from Bambang S. Yudhoyono, the President of Indonesian Republic, and in 2010, she retired from teaching mathematics. She started in 1987 at STKIP PGRI Banjarmasin as a temporary lecturer, and since in 2010 until now, she is a permanent lecturer at STKIP.

She has published many articles about education in the local newspaper, Banjarmasin Post. She presented articles about metacognition of mathematics education to the national conference mathematics in

Surabaya, Malang, Yogyakarta, and Banjarmasin. Her research about teaching and learning mathematics in primary schools has been recognized as one of the 20 best research of Indonesian Widyaiswara and got a reward from the Ministry of Education in 2008. Zahra loves singing and enjoys all kind sports, particularly table tennis. She lives in Banjarmasin with four children and eight grandchildren.

https://doi.org/10.1142/9789813146785_0005

Chapter V

IRAN: Mathematics Education in Iran from Ancient to Modern

Yahya Tabesh
Sharif University of Technology

Shima Salehi
Stanford University

Introduction

Land of Persia was a cradle of science in ancient times and moved to the modern Iran through historical ups and downs. Iran has been a land of prominent, influential figures in science, arts and literature. It is a country whose impact on the global civilization has permeated centuries. Persian scientists contributed to the understanding of nature, medicine, mathematics, and philosophy, and the unparalleled names of Ferdowsi, Rumi, Rhazes, al-Biruni, al-Khwarizmi and Avicenna attest to the fact that Iran has been perpetually a land of science, knowledge and conscience in which cleverness grows and talent develops. There are considerable advances through education and training in various fields of sciences and mathematics in the ancient and medieval eras to the modern and post-modern periods.

In this survey, we will present advancements of mathematics and math education in Iran from ancient Persia and the Islamic Golden Age toward modern and post-modern eras; we will have some conclusions and remarks in the epilogue.

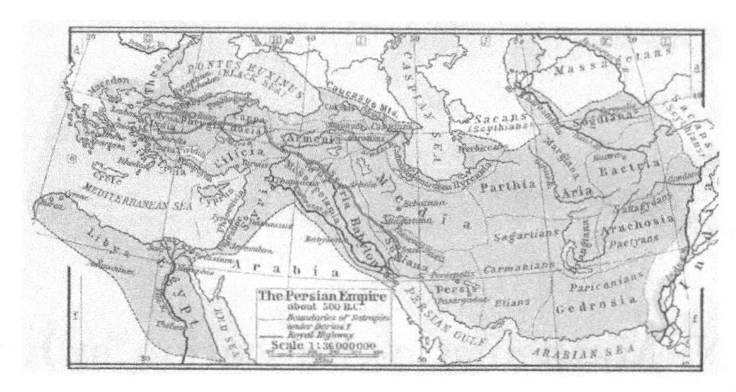

Persian Empire

Ancient: Rise of the Persian Empire

The Land of Persia is home to one of the world's oldest civilizations, beginning with its formation in 3200–2800 BC and reaching its pinnacle of its power during the Achaemenid Empire. Founded by Cyrus the Great in 550 BC, the Achaemenid Empire at its greatest extent comprised major portions of the ancient world.

Governing such a vast empire was, for sure, in need of financial and administration systems. In the 1930s, researchers from the Oriental Institute of the Chicago University found more than 30,000 clay tablets in Persepolis (capital of the Achaemenid Empire). Tablets provided information about the finance and administration systems, and they utilized weights and enumeration models, arithmetic and accounting systems. Besides arithmetic and numerical systems, geometry also had been used extensively in architecture to construct buildings and cities such as Persepolis. There is no evidence about any educational system or educational institute in the empire, but seems there had been some hands-on and in-service trainings.

In the ancient period, the only evidence of a structured educational state was in Gondishapur Academy during the Sasanian dynasty (224–650 AC). The Academy of Gondishapur was one of the most important learning institutes in the ancient time. It was the intellectual center of the Sasanian Empire and offered training in medicine, philosophy, theology and science. The faculty was versed in the Zoroastrian and Persian traditions. No evidence of mathematics or math education has been remained but there are some references to a "Royal Astronomical Tables" which indicates that reasonable amounts of math had been developed to provide such astronomical tables.

Medieval: The Golden Age

Muslim conquest of Persia led to the end of Sasanian Empire in 651, but after two centuries of the Arab rule, semi-independent and independent Iranian kingdoms began to appear, and by the Samanid era in the 9th and 10th centuries, the efforts of Iranians to regain their independence had been well solidified.

The blossoming literature, philosophy, medicine, arts and sciences of Iran became major elements in the formation of a new age for the Iranian civilization, during the period known as the *Islamic Golden Age*. The Islamic Golden Age reached its peak by the 10th and 11th centuries, during which Iran was the main theater of the scientific activities. After the 10th century, the Persian language, alongside Arabic, was used for the scientific, philosophical, historical, musical, and medical works, where the important Iranian scientists such as Tusi, Avicenna, and al-Biruni, had major contributions. Mathematics was also in the center of scientific development in the Golden Age and many original math works developed accordingly.

The history of mathematics during the Golden Age of Islam, especially during the 9th and 10th centuries, built on Greek mathematics and Indian mathematics but saw important development, such as full development of the decimal place-value system to include decimal fractions and the first systemized study of algebra, as well as advances in geometry and trigonometry.

One major contribution came from Mohammad ibn Musa al-Khwarizmi, who played a significant role in the development of algebra, algorithms, and Hindu-Arabic numerals in the 9th century. Al-Khwarizmi's contributions to mathematics, geography, astronomy, and cartography established the basis for innovation in algebra and trigonometry. His systematic approach to solving linear and quadratic equations led to algebra, a word derived from title of his 830th book on the subject, *The Compendious Book on Calculation by Completion and Balancing.*

On the Calculation with Hindus Numerals written circa 825, was principally responsible for spreading the Hindu-Arabic numeral system throughout the Middle East and Europe. It was translated into Latin as *Algoritmi de Numero Indorum*. Al-Khwarizmi rendered as (Latin) *Algoritmi,* which led to term "*algorithm*" in computer science.

A page from al-Khwarizmi's Algebra

Al-Khwarizmi provided exhaustive account of solving polynomial equations up to the second degree and discussed the fundamental methods of "reduction" and "balancing," referring to the transposition of terms to the other side of an equation, that is, the cancellation of like terms on opposite sides of an equation.

Al-Khwarizmi's method of solving linear and quadratic equation worked by first reducing the equation to one of the six standard forms and by dividing out the coefficient of the square. By using the two operation *al-jabr* ("restoring" or "completion") and *al-muqabala* ("balancing"), he developed a methodology for solving equations.

Another famous mathematician of the 11th century is Omar Khayyam, who wrote the influential book *Treatise on Demonstration of Problems of Algebra* (1070), which laid down the principles of algebra and derived general methods for solving cubic equations and even some higher orders.

In the *Treatise*, he wrote of the triangular arrays of binomial coefficients also known as Pascal's triangle. In 1077, Khayyam wrote *Sharh ma Ashkala min Musadarat Kitab Uqlidis* (Explanations of the

Difficulties in the Postulates of Euclid) published in English as "On the Difficulties of Euclid's Definitions." An important part of the book is concerned with Euclid's famous parallel postulate, and Khayyam's advances and criticisms may have contributed to the eventual development of non-Euclidean geometry.

As a mathematician, Khayyam has also made fundamental contributions to the philosophy of mathematics especially in the context of Persian Mathematics and Persian philosophy.

Another significant Persian mathematician and astronomer in the medieval period of the 14th century was Ghiyath al-Din Jamshid Mas'ud Al-Kashi. Al-Kashi developed great contributions in numerical systems and approximation methods. In his famous book *The Treatise on the Chord and Sine*, Al-Kashi computed sin 1° to nearly as much accuracy as his value for π, which was the most accurate approximation of sin 1° in his time and was not surpassed until 16th century. In algebra and numerical analysis, he correctly computed 2π to nine sexagesimal digits, and he converted his approximation of π to 17 decimal places of accuracy. Al-Kashi also developed and used both decimal and sexagesimal fractions with great ease in his *Key to Arithmetic*.

Al-Kashi's work included astronomy as well. He wrote the *Treatise on Astronomical Observational Instruments*, which described a variety of different instruments, and he invented the *Plate of Conjunctions*, an analogue computing instrument used to determine the time of day by using linear interpolation techniques. Al-Kashi also invented a mechanical planetary computer which he called the *Plate of Zones*, which could graphically solve a number of planetary problems, including the prediction of the true positions in longitude of the sun, moon, and planets in terms of elliptical orbits, the latitude of the sun, moon, and planets, and the ecliptic orbit of the Sun.

In the Golden Age, for scholars training including mathematicians, there existed scientific institutes. In 832, the famous House of Wisdom was founded in Baghdad, there the methods of Gondishapur were emulated and many mathematicians and scientists attracted and joined. Some scholars produced original research in the House of Wisdom for example, al-Khwarizmi worked in House of Wisdom for development of his contribution on algebra. In the northern and central parts of Iran there were also libraries under support of Persian kingdoms where intellectual and scholar training centers and many important scientists and mathematicians were housed and supported. There is no evidence of any educational system or institute other than scholarly training under supervision of scientists and mathematicians.

In the eleventh century Nezamiyeh Institutes were founded as a group of the medieval institutions of higher education by Khwaja Nezam al-Mulk (the prime minister in the Sejukian era, the title of *Nezamiyeh* derived also from his name). Nezamyieh institutes were among the first well organized institutions of higher learning in Iran. The quality of education was among the higher in the Islamic world, and they were even renowned in Europe. The royal establishments of the Seljukian Empire and elite class supported them financially, politically, and spiritually. The Nezamiyeh Institutes placed more emphasis on religious studies but science, mathematics and astronomy also were considered.

In 1219, Iran suffered a devastating invasion by the Mongol army of Genghis Khan. The Mongols violence and depredations killed up to three-fourths of the Iran's population of the Iranian, as libraries were burned and a civilization was destroyed. Following the demise of the Mongol Empire in 1256, Hulagu Khan, grandson of Genghis Khan, established the Ilkhanate in Iran. In 1370, yet another conqueror, Timur, followed the example of Hulagu, establishing the Timurid Empire that lasted for another century and half. The Ilkhans and Timurids soon came to adopt the customs of the Iranians, choosing surround themselves with a culture that was distinctively the customs of the Iranian. Nevertheless, the scientific establishments never reached to the pre-Mongolian in Iran.

Modern Era: Post Mongolian

In the 1500s the Safavi Empire reunited Iran, and European industrial products touched the Iranian market, but for more than three centuries, no approach for any reform in educational systems could be seen. Theological education dominated the traditional schools, which caused math and science education to hold little value.

Three centuries later, the Russo-Persian war ended in 1828 by the Turkmenchay treaty. Russia completed the conquering of all Caucasian territories from Iran, having previously gained Georgia, Dagestan, and most of contemporary Azerbaijan through the treaty of Gulistan in 1813. After losing the wars, the government officials understood that military losses were due to a lack of modern science and technology, so they dispatched a group of students to Europe to be trained in modern sciences. The next top down approached was performed by Amir Kabir, the reformist prime minister of the mid-19th century. He established Dar ul-Funun in 1851 as the first modern higher education institute in Iran. Emphasis was placed on medicine, science and military training, but a modern approach to math education was also in agenda. The public educational system was limited to the traditional elementary schools with emphasis on Arabic and Persian literatures and without much focus on mathematics teaching just arithmetic and a special accounting system called *Siaq*. The next bottom up approach was made in 1887 by Mirza Hassan Roshdiyeh who founded the first modern elementary school in Tabriz in the northwest of Iran. Later in 1897 he established the first modern elementary school in the capital Tehran, but there was still no establishment for the inclusive modern educational system.

In 1906, the constitute revolution caused more new approaches to be made toward reforming the traditional systems. In 1909, the second parliament passed a bill to establish the Ministry of Maearif (culture and science) and to officially establish educational systems including elementary and high schools and also higher education. As a result, elementary and high schools were founded nationwide. Teaching and learning mathematics were core value of the new system, to train more needed teachers including math teachers, the Teachers College was established in 1920.

Teachers College trained professional math teachers; the curriculum included psychological and teaching skills too.

In 1921, the parliament passed another bill to dispatch 100 students annually to Europe for five years to be trained in modern science and technology which included mathematics and math education. Some educated mathematicians and math educators became pioneering leaders in math education reform by developing new curriculums and writing math textbooks at all levels.

In 1934 Tehran University was established as the first modern higher education institute in Iran. Some smaller independent colleges also joined the new university including the Teachers College. Tehran University trained many professional mathematicians and also math teachers.

In 1941 Iran was occupied by the Allies troops during World War II, but after the war ended in 1948 new universities in major cities as Mashhad, Tabriz, Isfahan, and Ahvaz established. Mathematics came to be considered one of the most important majors, and many math teachers taught nationwide. In the 1960s new universities as Aryamehr (Sharif) University of Technology and Pahlavi (Shiraz) University were founded. New universities reformed the educational system, creating new approaches in the math curriculum that impacted the math curriculum in the pre-university level as well.

The pre-university educational system started with six years of elementary and six years of secondary schools. In elementary school, arithmetic and basic concepts of geometry were taught, in the secondary level subjects such as algebra, geometry, pre-calculus and calculus were covered. In algebra and pre-calculus and calculus, courses were primarily based on numerical and symbolic calculation, but in geometry logical

Teachers College (Tehran).

thinking and deductive reasoning were in agenda. In the 1960s the educational system was reformed to comprise five years of elementary, three years of middle and four years of high school. The worldwide Modern Math approach in the 1960s also impacted the math curriculum, causing subjects such as set theory and logic to appear in textbooks.

The Iranian Mathematical Society was also established in the late 1960s and gathered mathematicians as well as math teachers in annual conferences to discuss future advancements in mathematics.

21st Century: Post Modern Era

In this 21st century the youth majority of the Iranian population is eager for progress and looking for empowerment in the knowledge-based economy. More than 5% of Iran's population of 80 million are college students, and almost half of them are girls. The number of K–12 students reached 20 million students in the early 2000, but it has come down in 2015 to 12 million students, almost 15% of the population. Mathematical literacy is at the heart of the educational system, both in K–12 and college level. Some highlights about the educational system and math education follow.

- The Ministry of Science, Research and Technology is responsible for higher education and research as well as technological development. The Ministry of Education is responsible for K–12 education.
- There are over 100 public universities and 300 private colleges in Iran. Also, Azad private university and vocational colleges are established in most of major cities. Four and half million students lie at the college level. There are almost 100,000 math major students in undergrad and graduate levels.
- Teachers University and most major universities offering a math teaching degree in the undergrad level. Math education programs in the MSc and PhD levels are also offered.
- PhD programs in mathematical sciences and math education since late 1980s have institutionalized research in math and math education. Research has been supported widely by the public

research institutes. More than 5,000 research papers have been published in the period of 2008 to 2014 in the fields of mathematical sciences.

- The Institute for Research in Fundamental Sciences (previously the Institute for Studies in Theoretical Physics and Mathematics, often shortened to IPM) established in the late 1980s and is a major sponsor of research in mathematical sciences nationwide.
- The Iranian Mathematical Society, as a professional association is supporting math and math education through conferences, workshops and seminars also publications. The Annual Math Conference over the last five decades and biyearly Math Education Conference are the most important conferences. The *Bulletin* of the Iranian Mathematical Society, the most important research journal, *Farahang-o-Andisheh Riazi* the expository journal and a monthly newsletter are the most important publications of the society. Iranian Mathematical Society also organizes a yearly student conference and math competition among the undergraduate math students, each university can participate with a team of five students in the competition and top teams are supported to participate in the International University Students Math Contest.
- The Iranian Statistical Society, also established in 1990. The society is organizing conferences and publishing journals for scientists and professionals.
- The Iranian Math Teachers Association is a nationwide professional association with most math teachers as members in the local chapters. Local chapters organizing seminars and workshops and publishing journals. Association also is co-sponsor of the biyearly math education conference which is the largest gathering of math teachers in Iran.
- There are more than one million teachers in more than 100,000 schools nationwide, of which more than 90% are public schools.
- The national curriculum is in the form of six years of elementary and six years of high school. Math for all students constitute three to five hours of each school week up to grade 9. Specific branches are taught in grades 10 to 12: math-physics, natural sciences,

humanities and vocational training. 10 to 12 hours of math courses are taught per week for math-physics and three to five hours for others.

- Number systems, arithmetic and basic geometry are covered in the elementary level, while algebra, pre-calculus, calculus, combinatorics, probability and geometry are covered in high school.
- Standard textbooks at the national level are provided and published by the ministry of education at over 120 million copies per year. The private publishers publish enrichment books.
- Online and digital learning systems are currently progressing rapidly; *maktabkhoneh.org* and *kelasedars.org* are some of the most popular for math courses.
- The National Organization for Development of Exceptional Talents (NODET) was established in 1976 to support gifted students in special schools in the middle school and high school levels nationwide. NODET is supporting math education through advanced enrichment courses.
- The Iranian Mathematical Olympiad is a math contest for high school students organized since 1985 by the Young Scholars Club under the support of the Ministry of Education. Iran's team participated in the International Math Olympiad for the first time in 1987 in Cuba and received just one Bronze medal, but math contest at the high school level quickly became very popular among high school students. Nowadays, with almost 100,000 applicants annually, through three rounds of contests the national team of six students is selected to participate in the International Math Olympiad. In the last three decades, they usually end up in top ten and ranked first in 1998 with five gold medals and one silver medal. The Math Olympiad as an extracurricular activity has attracted many talented students to mathematical studies which empowered them beyond the standard curriculum.
- Math journals have also had a deep influence in the popularization of mathematics and extracurricular math education. The math journal called *Yekan* started publishing in the 1960s and for two

decades it was the most popular journal. Recently *Roshde Amoozesh Riazi,* and *Nashri Riazi* and many other local and nationwide math journals have also attracted many interested students and math teachers.

Isfahan Mathematics House

- Math House is a unique community center for extracurricular activities in mathematics for students. Empowering and attracting youth to mathematical sciences are primary objectives of Math House. Students are attracted to participate in workshops and seminars and join project groups to develop basic and new concepts in group projects to present in Math Fests quarterly. The first math house was established in Isfahan in the central part of the country in year 2000 on the occasion of the World Mathematical Year-2000. Now, more than 30 math houses are established in major cities.
- The mathematics movement in Iran is flourishing in the 21st century and Iranian mathematicians are influencing the world by developing mathematical sciences. These achievements are reflected in various ways, such as the most prestigious Fields Medal awarded to Maryam Mirzakhani, the first women to ever earn the medal. Maryam did her middle school and high school studies in Iran and, as a gold medalist in the International Math Olympiad finished her undergrad in mathematics in the Sharif University of Technology in Iran. She later received her PhD from Harvard University and served as a professional mathematician.

Epilogue

In the long journey that is Iranian math education, from the ancient period to the postmodern era, there are many lessons that can be learned but to look forward for further advancement of math education, we can look at some obstacles and how to overcome them.

There are two main obstacles in the Iranian educational system as well as math education. First the system is centralized and non-flexible. Second, the system is heavily test oriented and focused on memorization, which can hinder students' creativity.

We may consider a vision of a knowledge-based society and may consider the following issues to reform math education and the educational system as a whole:

- Decentralization, flexibility and diversity
- Devolution of decision power and responsibility to the local level
- Commitment to a culture of trust through professionalism
- Equal educational opportunity
- Look for broad knowledge as new skills are needed
- Emphasizing problem solving in math education
- Developing an experimental learning system in math learning
- Pushing a system that courage learners to be active partners and just not passive receivers
- Developing a project based and inquiry based learning system
- Developing a systematic, flexible and individual support system
- Technology should be used as an enabler for interactive and personalized learning

The motto of the educational system in Iran is best highlighted by Iranian poet Ferdowsi, who says "knowledge is power." Let us hope this education gives power for peace and friendship all over the world.

Knowledge is power.

References

[1] Cameron, George C., **Persepolis Treasury Tablets**, The University of Chicago Press 1948.

[2] Fisher, William B. et al., **The Cambridge History of Iran**, Cambridge University Press 1989.

[3] Taghizadeh, S. Hassan, **History of Science in the Golden Age**, Tehran University Press 1957.

[4] Nasr, S. Hossein, **Science and Civilization in Islam**, ABC International Groups Inc. 2007.

[5] Schlegel, Flavia et al., **UNESCO SCIENCE REPORT, Toward 2030**, UNESCO 2015.

[6] **Iran Statistical Report**, Iran Stat Organization 2014.

About the Authors

Yahya Tabesh is a visiting scholar at Stanford University and distinguished faculty member of the Sharif University of Technology (Tehran, Iran). Yahya served as the chairman of the department of Mathematical Sciences at Sharif for several years. He was also director of Math and Computer Olympiads in Iran. A pioneer in developing math house and schoolnet, he is also responsible for developing mathematics textbooks and curriculum development for high school students as a member of the high council of educational reform in Iran. Dr. Tabesh also served as director of the Computing Center at Sharif University. He also did research in computational linguistics and was a leader in developing Farsi under the Unicode standard. Dr. Tabesh used to be member of the high council of informatics in Iran. He won the Erdós International Award in 2010 for his sustained and distinguished contribution to the enrichment of math education. Dr. Tabesh currently is working on cognitive learning system to develop an online interactive learning engine.

Shima Salehi is a PhD student in the Learning Sciences program at Stanford Graduate School of Education. She has a B.Sc. in Electrical Engineering and a M.A. in Learning, Design, and Technology. Her research focus is at the intersection of science education, technology, and equity. Shima is interested in examining how scientific competency could be promoted in students, and how technology and socioemotional factors could either hinder or foster this process. In her previous work, she has examined the effect of computational modeling on students' scientific competency. She has studied the variations in educational technology use across genders and racial groups. Shima has also studied methods for optimally integrating interactive simulations into instructional practices.

IRAN: Appendix

Arithmetic 1st grade 1968

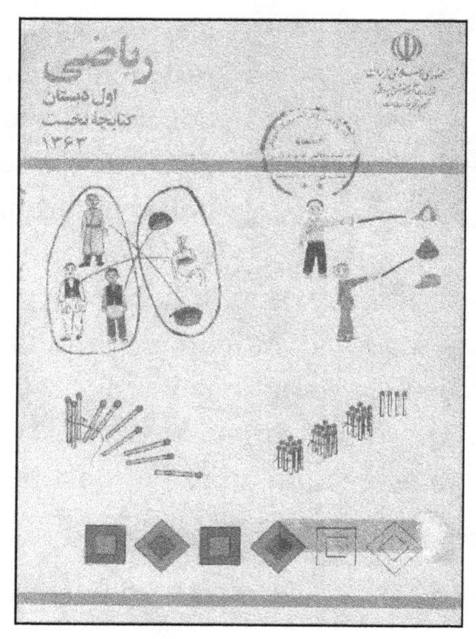

Mathematics 1st grade 1984

Mathematics 1st grade 2015

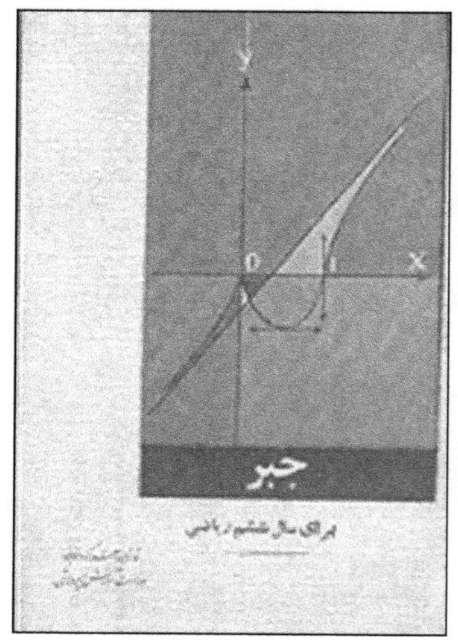

Algebra 12th grade 1968

Modern Math 9th grade 1980

Discrete Math 12th grade 2015

Yekan, math journal 1963

Nashri Riazi, math expository journal 1986

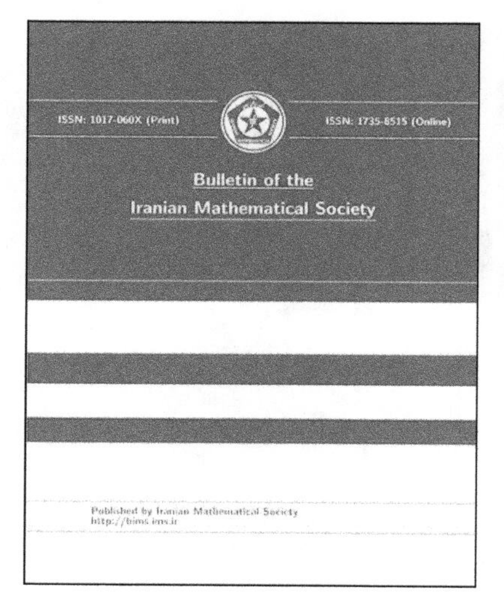

Bulletin IMS, math research journal 2016

Calculus by: S. Shahshahani, National Book Award 2010

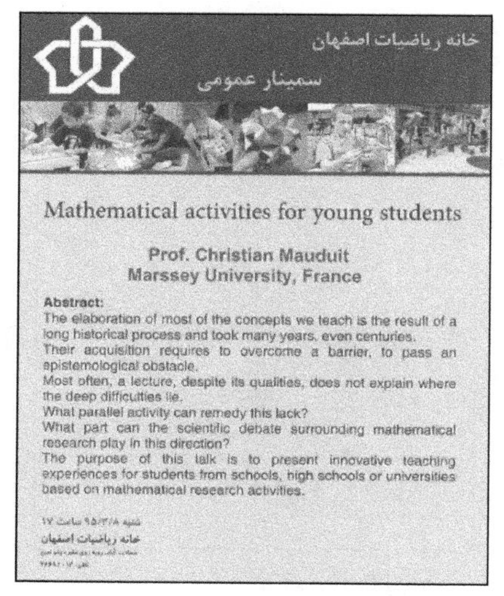

Isfahan Math House, a learning environment 2016

47th Annual Math Conference 2016

https://doi.org/10.1142/9789813146785_0006

Chapter VI

JORDAN: Mathematics Education in the Hashemite Kingdom of Jordan

Rana Alabed
The Jubilee School

Ma'moun El-Ma'aita
Ministry of Education, Jordan

The Hashemite Kingdom of Jordan is an Arab Muslim country, located in the north of the Arabian Peninsula and in West Asia. Bordered by Syria to the north, Iraq to the east, Saudi Arabia to the south and south-east, and Palestine (the West Bank) to the west. Jordan is named after the Jordan River, which passes on its western border. Amman is Jordan's capital city and the seat of the King and government. The official language is Arabic, and English is the first foreign language.

The Hashemite Kingdom of Jordan is a hereditary monarchy. Majesty King Abdullah II sits on the throne of the Kingdom and supervises the top three authorities, as well as serving as the supreme commander of the armed forces.

People of Jordan

The population in 2011 was 6,249,000. Arabs make up the vast majority (98%) of the population, and Jordan's other ethnicities include Alcharkas (circassians) (1%) and Armenians (1%). Population density is concentrated in the center and north of the country. The official religion is Islam, as Sunni Muslims comprise 92% of the population. Other Islamic sects such as the Druze make up just 2% of the population. Christians comprise 6% of the population, most of whom follow the Orthodox Church.

Area and Climate

The total area of the Kingdom is 92.300 square kilometers. The total length of 1,635 km border includes 744 km shared with Saudi Arabia, 375 km with Syria, 238 km with the 1948 Green Line, 181 km with Iraq, and 97 km with the West Bank. Jordan has a port on the Red Sea through the city of Aqaba, located in the far north of the Gulf of Aqaba. The lowest point is the surface of the Dead Sea and is -408 m below sea level. The highest point stands at 1854 m on the summit of Mount Umm Al-Dami.

Business Practices

Following Islamic practice, government departments and offices, banks and most of its other offices close on Friday and Saturday of each week. The working hours on the remaining five days (Sunday–Thursday) are from 8:30am to 3:30pm.

Education in Jordan

Introduction

The Ministry of Education has been seeking to achieve educational development that is based on genuine national and educational principles. The Ministry was able to achieve different accomplishments at various levels and aspects. Inspired by the Hashemite Leadership insightful visions that foresee a future holding prosperity for Jordan, the Ministry of Education has been enriching both knowledge and learning streams by providing qualified educational experiences. To proceed with this process, the Ministry has adopted research and planning approaches to meet clear objectives for the purpose of building a knowledge-based society. Consequently, many fruitful and crucial achievements were accomplished.

Historical Overview of the Ministry of Education

Jordan provides an exemplary educational system which is capable of achieving distinguished progress in education, in cultural accomplishments and successful national experiences that put Jordan at a front rank amongst the developed countries in this field.

The educational system seeks to prepare students and qualify them to become more active in a community that moves towards a "Knowledge Economy." It also gives much importance to achieving sustainable development through human resources development to enable Jordan to become a hub for Information and Communications Technology (ICT) in the region. The aforementioned progress would not have been achieved without true commitment, dedication and firm planning and support of a judicious leadership.

Education in Jordan has gone through several phases since the founding of the Kingdom, culminating in the reign of the late King Hussein ibn Talal.

His Majesty King Hussein gave education utmost importance when he assumed his constitutional powers. The First National Conference for Educational Development was held on September, 1987 and affirmed the following principles:

- Emphasizing the importance of political education in the educational system and enhancing the principles of participation and justice.
- Providing for continuous education.
- Focusing on the development of an individual's capabilities for analysis, criticism and initiative taking.
- Enhancing scientific methodology in the educational system and developing research skills.
- Ensuring centralization in planning and follow-up, and decentralization in administration.

Educational phases were also approved in the same conference as follows:

1. Kindergartens: A two-year stage.
2. Basic education: A ten-year stage.

3. Secondary education: A two-year stage, consisting of two tracks:
   - Comprehensive secondary education with a common general education basis and a specialized academic or vocational education.
   - Applied secondary education which is based on vocational preparation

Education during His Majesty King Abdullah II ibn Al Hussein's Reign

As soon as His Majesty King Abdullah II ibn Al Hussein assumed his constitutional powers, upon the death of his father King Hussein, education gained a great deal of his attention and patronage. He strongly believed in the importance of building a knowledge capital in the socio-economic development process of the nation. His Majesty focused on the principles of equality and equity in opportunities and on providing standardized educational services throughout the Kingdom.

His insight and vision appeared clearly in the appointment of the successive Jordanian governments that stressed the importance of developing education through various interventions focusing on educational quality, the appropriate deployment of information and communication technology, the expansion in pre-school education and providing the suitable learning environment and necessary programs to achieve professional development for teachers.

His Majesty emphasized the importance of transforming Jordan into a "Knowledge Economy" through involvement in the international knowledge economy, and meeting its challenges in order to prepare Jordan to become a center for information technology, and an advanced example in educational achievements that are based in knowledge economy. The Ministry of Education exerted considerable efforts within this vision to develop students' scientific research skills through highly developed curricula and moved forward in disseminating kindergarten throughout the kingdom.

The Ministry of Education adopted the first phase of "Education Reform for Knowledge Economy" project (ERfKE I) in the Vision Forum for the Future of Education in Jordan in 2002. This project achieves results that put Jordan in a leading position among the Arabs and the rest of world. The Ministry has concluded the second phase of "Education Reform for Knowledge Economy" project (ERfKE II) for the years 2009–2013 that focused on deepening the qualitative effect of education and on accrediting schools as basic units for development.

The Ministry expresses this concept as follows: "A human being is the most precious resource we have" and that the teacher is the leader of the teaching-learning process who educates generations and helps them achieve excellence and innovation. In this regard, His Majesty's initiative came to provide suitable teaching-learning environments for teachers and students, which reflected the political will for development. His Majesty was keen on directing royal letters to the students and the education staff at the beginning of each scholastic year. In order to enhance the teaching-learning process His Majesty decreed:

1. Exempting students from tuition fees.
2. Distributing winter coats to students and heating up schools.
3. Expanding school nutrition project including (UNRWA) schools.
4. Providing full health and nutritional care to students.
5. Providing lifelong learning to those who are beyond the schooling age so that they can keep pace with the rapid social and technological changes and enable them to support a good standard of living.
6. Developing under-privileged areas.
7. Activating King Abdullah II's Prize for Physical Fitness.
8. Establishing King Abdullah II's Schools for Excellence.
9. Establishing teachers' clubs throughout the Kingdom.

Mathematics in the Jordanian Curriculum

The several mathematics applications in the various fields of science, such as engineering, economics, agriculture, basic sciences and humanities, gave mathematics a leading role in the evolution of human civilization in the past and the present. In the Kingdom of Jordan mathematics will continue to play this role. Man needs mathematics in his life; to perform calculations, analyze data, communicate with others, solve problems, and make decisions.

The technology that is primarily based on mathematics, has accelerated changes in our world, and influenced the way we live, teach, and learn. Because of this rapid change, students need to learn math to help them in their future, and allow them to develop their problem-solving skills, decision-making, rational justification in their daily lives, and to apply the principles and skills they need in the technology community, provided that they learn it in ways that motivate them to continue learning independently.

The contemporary global trends emphasize to direct education towards the development of a national economy based on knowledge, skills and experience needed for students, to enable them use it in practical life and to harness technology for the production of knowledge, transferring and sharing it, for the development of society.

The application of the principles of mathematics in problem solving help students develop the knowledge and skills they need in a technological society.

The use of technology has significantly changed the teacher's role, as the teacher no longer is the only source of knowledge, students are taught using information and communication technology tools, to evaluate the solutions to the problems they face, and to analyze large amounts of data, using tables and databases.

These tools help students learn meaningful mathematics, and develop an in-depth understanding of them. The calculators and computer software should play an important role in mathematics teaching and learning, as catalyst for learning, since they remove the burden of calculations that accompany the solution to the problem at hand, these tools allow using actual data-related to certain issues, and can be used to develop concepts, support and encourage the process estimators, and help to survey and detect sequence processes.

The development of mathematics curriculum has become an urgent need, in view university professors, studies related to curriculum evaluation, and field observations made by (supervisors, teachers, students, parents and the local community) about the curriculum, and the results of national and international tests that showed some of the weaknesses in it, the development of the curriculum will prepare students better for the future.

Considering evaluation strategies and the diversity in applying it in the classroom, will activate students learning process, and will shift teaching from teacher-centered teaching, to the learner-centered teaching.

The key learning outputs from the study of mathematics include:

Students should be able to:

- Appreciate the role of mathematics in improving the quality of life of individuals and society.
- Connecting mathematical ideas and applications in Islamic Arab culture.
- Accept the ideas of others and discuss their solutions while working with them, and provide feedback.
- Demonstrate confidence, perseverance, honesty, and cooperation by individuals when learning and applying mathematics.
- Appreciate the role of mathematics in building positive human relations between the cultures of the world, as a universal language evolved from different civilizations.
- Recruitment thinking skills for lifelong learning skills.
- Functioning data processing to get inferences and predictions.
- Develop capacity in the logical justification for learning mathematics independently, and through working with others, and contribute positively as a leader or a member of the team.
- To use of technology tools such as: (software, calculators, and computers) effectively to understand mathematics in depth.
- To choose the suitable methods or tools (mental arithmetic, estimation, and pen and paper, and computers) when making calculations.
- Apply mathematical skills, processes and modeled, efficiently and accurately in everyday life.
- Use problem-solving to generate mathematical knowledge.
- Establishing integration between mathematics and other sciences.
- Appreciate the role of the world scientists, and especially Arabs and Muslims, who have contributed in developing mathematics.

These desired outcomes have guided the development of Jordan's national mathematics curriculum through each of the three stages of national education.

Table 1. Main pivotal mathematics through Jordanian Curriculum.

Cycle	Pivot
Basic education Grade (1–8)	• Numbers and Operations • Patterns and Algebra • Measurement • Geometry • Statistics and Probability
Basic education Grade (9–10)	• Numbers and Operations • Algebra (Leaner programming, Polynomials) • Geometry • Measurement • Statistics and Probability
Secondary education Grade 11 Scientific stream	• Algebra: Functions (Real, Exponential, logarithmic, trigonometric), equations, inequalities, triangles, sequences, and series. • Statistics and Probability
Secondary education Grade 11 Literature, Shariee, Nursing, Industrial, and Hostelling stream	• Algebra sequences, and series: Functions (Exponential, logarithmic), Rational Functions • Statistics and Probability
Secondary education Grade 12 Scientific stream	• Algebra: conic sections, Limits, Continuity, calculus: Differentiation and its applications, Integration and its applications. • Statistics and Probability
Secondary education Grade 12 Literature stream	• Algebra: Limits, Continuity, calculus: Differentiation and Integration and some applications • Statistics and Probability

This curriculum continues to undergo evaluation and adaption to the capabilities of teachers and the needs of teachers.

School Schedules

The schools in Jordan consist of 190 days where students in basic education attend school five days a week. Generally, students attend one mathematics class per day (45 minutes for one period). The table below shows the number of mathematics class periods per grade per year. In addition to these classes, students participate in review classes and take exams and quizzes outside the regular class times.

Table 2.

Education	Grade	Stream	Number of classroom periods
Basic education	1	---	124
	2	---	117
	3	---	125
	4	---	143
	5	---	153
	6	---	165
	7	---	149
	8	---	166
	9	---	172
	10	---	164

In secondary education the class period is one hour distributed as follows:

Education	Grade	Stream	Number of classroom periods
Secondary education	11	Scientific Stream	121
	11	Literature, Shariee, Nursing, Industrial, and Hostelling	68
	12	Scientific Stream	125
	12	Literature, Shariee, Nursing, Industrial, and Hostelling	92

Table 3. The General Outputs for basic education grade (1–8).

Pivot	Outcomes
Numbers and Operations	• To understand numbers and relations between it. • Understand the meaning of the four arithmetic operations • Do calculations in the four arithmetic operations and give reasonable estimates.
Algebra (patterns)	• To understand patterns, relations, functions. • To use symbols and algebraic expression and equations and inequalities in mathematical issues. • To use and understand the mathematical models to express quantitative relations.
Geometry	• To analyze the characteristic of the geometric shapes in two and three dimensions, and make mathematical justifications about geometric relations. • Apply geometric transformation, and using symmetric to analyses mathematical positions • To use visual and spatial inference and geometric models in problem solving.
Measurement	• To understand measurable shapes, measurement systems, and the operations about them. • Applying technology and apparatus for measurement.
Statistics and Probability	• Collecting data, organize it, and present it. • Analyzing a statistical data. • Conducting assumptions and making decisions. • Understanding probability and applying its main concepts.

Table 4. The General Outputs for basic education grade (9–10).

Pivot	Outcomes
Algebra (patterns)	• To understand patterns, equations, and functions and applying it in problem solving. • To use symbols and algebraic expression and equations and inequalities in mathematical issues. • To use and understand the mathematical models to express quantitative relations. • To analyses change in various situations.
Geometry	• To analyze the characteristic of the geometric shapes in two and three dimensions, and make mathematical justifications about geometric relations. • Apply geometric transformation, and using symmetric to analyses mathematical positions • To use visual and spatial inference and geometric models in problem solving.
Measurement	• To understand measurable shapes, measurement systems, and the operations about them. • Applying technology and apparatus for measurement.
Numbers and Operations	• To estimate some arithmetic operations reasonably. • To solve problems contains variety of economical applications.
Statistics and Probability	• Collecting data, organize it, and present it. • Analyzing a statistical data. • Conducting assumptions and making decisions. • Understanding probability and applying its main concepts.

Table 5. The General Outputs for secondary education scientific stream grade (11–12).

Pivot	Outcomes
Algebra: Functions (Real, Exponential, logarithmic, trigonometric), equations, inequalities, triangles, sequences, and series. conic sections, Limits, Continuity. Calculus: Differentiation and its applications, Integration and its applications. Statistics and Probability	• To understand patterns, equations, and functions and applying it in problem solving. • Use technology when graphing exponential and logarithmic functions. • To apply polynomials in problem solving. • To design a plan to collect data, analyze it and give predictions and statistical inferences. • Use differentiation concepts to solve life issues. • Use integration concepts to solve life issues. • To investigate the locus of conic sections and use it in problem solving. • To choose the best statistical way to analyze data.

Table 6. The General Outputs for secondary education Literature, Shari'eh, Nursing, Industrial, and Hostelling stream grade (11–12).

Pivot	Outcomes
Algebra sequences, and series: Functions (Exponential, logarithmic), Rational Functions. Limits, Continuity, calculus: Differentiation and Integration and some applications Statistics and Probability	• To understand patterns, equations, and functions and applying them in problem solving. • To understand special functions and to graph it using technology. • To apply polynomials in problem solving. • Use differentiation concepts to solve life issues. • Use integration concepts to solve life issues. • To design plan to collect data, analyze it and give predictions and statistical inferences. • To choose the best statistical way to analyze data.

Vision of Ministry of Education

The Hashemite Kingdom of Jordan has the quality competitive human resource system to provide all people with life-long learning experiences relevant to their current and future needs in order to respond to and stimulate sustained economic development through an educated population and a skilled workforce

Mission of Ministry of Education

To create and administer an education system based on "excellence," energized by its human resources, dedicated to high standards, social values, and a healthy spirit of competition, which contributes to the nation's wealth in a global "Knowledge Economy"

In the latest review of the curriculum of mathematics by the Ministry of Education, it emphasizes critical and creative thinking, and problem solving. This idea aligns with the Ministry of Education's Vision and Mission mentioned above.

Also emphasis was made on relating math to real life situations, through problem solving as well as through inter-disciplinary instruction in order to relate math to other disciplines.

The MoE also encourages scientific research by delegating students annually to INTEL ISEF competitions held in the U.S. following local short-listing. This award was initiated by INTEL and organized by (Science Service) which organizes and manages the International Science and Engineering Fair (ISEF). In 2013 one Jordanian student named Salah Eddin Ibrahim Abu Sheikh achieved the second place of the grand prize in the field of mathematics. He solved an equation that hasn't been solved before, and was awarded a prize by NASA by giving his name to a newly discovered minor planet. It was named ABU SHEIKH.

Jordan also has the annual Queen Rania Al-Abdullah Award for excellence. This award is offered to teachers, which encourages creativity and excellence among teachers.

References

1. www.jordan.gov.jo/wps/portal/!ut/p/b1/04_SjzQyMDc3MzW2NDfRj9CPykss y0xPLMnMz0vMAfGjzOLDLL0twrzdDQ3cA0yMDDy9LHzCzA1NjQ0CjfS DU_P0c6McFQFV6bWh/
2. http://www.moe.gov.jo/Directorates/DirectoratesSectionDetails.aspx?DirectoratesSectionDetailsID=137&DirectoratesID=36
3. http://www.moe.gov.jo/en/MenuDetails.aspx?MenuID=40
4. http://www.moe.gov.jo/en/MenuDetails.aspx?MenuID=41
5. Annual Report 2011, Ministry of Education Year Book 2012
6. General Framework and the General and Privet Outcomes of Mathematics, Basic and Secondary Education, Jordanian Curricula, second edition, 2013.

About the Authors

Rana Alabed holds a master's degree in Measurement and Evaluation from Yarmouk University, Jordan, and a bachelor's degree in Mathematics from the University of Jordan. Ms. Alabed worked with the Ministry of Education of Jordan as a high school math teacher (1980-1995) and was a member of the team who reviewed and revised math textbooks for the national curriculum.

Ms. Alabed started at the Jubilee School, King Hussein Foundation (a special school for the gifted and talented students in Jordan) in 1995 as a math teacher, and then was Head of the Math and

Computer Science Department for two years. Since 2001, she has been Head of Admissions and Registration Department at the Jubilee School for Gifted and Talented students, and at the same time she is working on the Battery of Developmental Assessment (BDA) for children. Ms. Alabed is also on the Board of the Royal Chess Union of Jordan and is the supervisor of the Math & Chess Program at the Jubilee Institute.

Ma'moun El-Ma'aita holds a Higher Diploma in Education from the Arab Open University, Jordan, and bachelor's degree in Mathematics and a minor in computer science from the University of Jordan.

Mr. Ma'aita started in the Ministry of Education of Jordan as a math teacher in high schools and community colleges, then he joined the programmers' team at the computer center in the Ministry of Education. With Mr. Ma'aita's knowledge and experience in programming, systems analysis, project management and evaluation, feasibility studies, and in the field of statistics, he was able to develop the educational statistics system in the Ministry of Education and also to contribute to education development & computer systems development in education.

Mr. Ma'aita was promoted in the Ministry of Education and served in many positions: the head of Exams and Registration department in the Training Directorate, the head of the Statistics department in the planning directorate, Director of Computer Systems & Data Bases, as well as being a consultant to the minister of education for technological issues.

JORDAN: Appendix

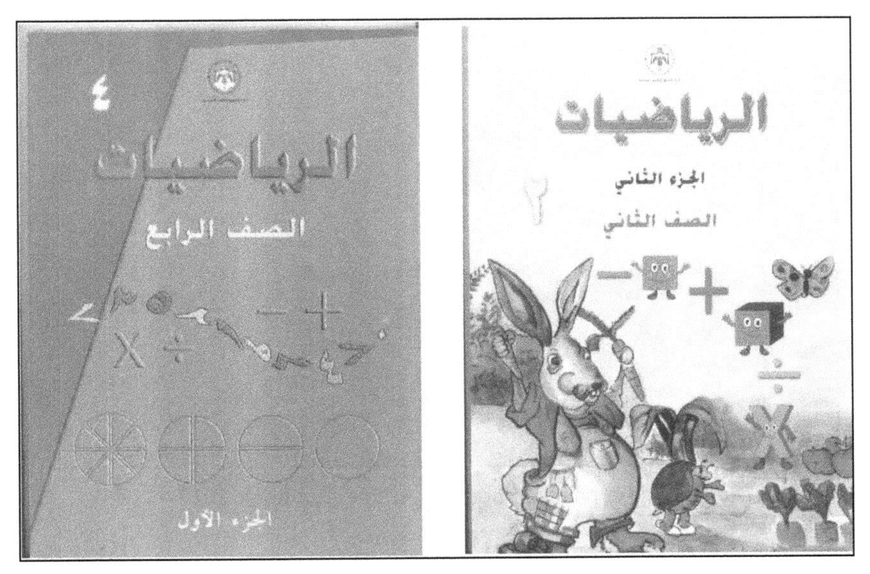

Left: Cover, 4th grade Arabic Textbook 2013
Right: Cover, 2nd grade Arabic Textbook 2013

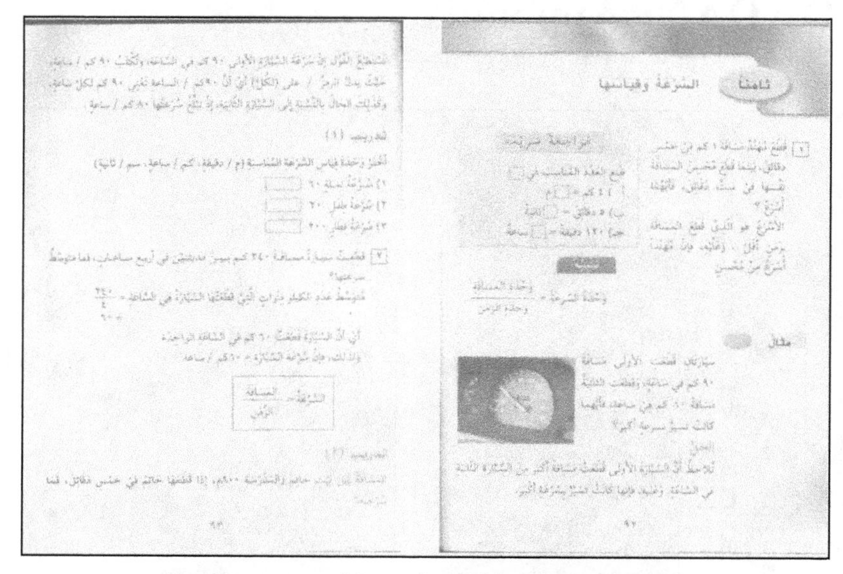

Random pages, 4th grade Arabic Textbook 2013

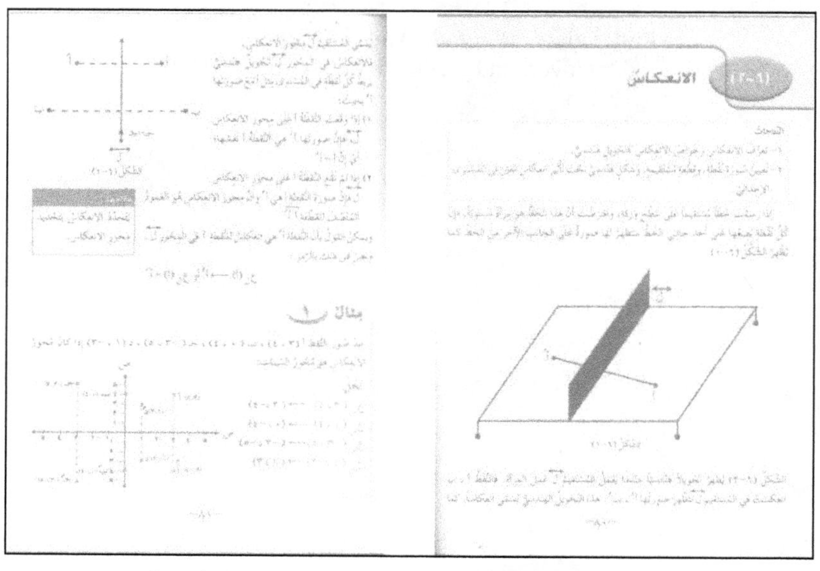

Random pages, 7th grade Arabic Textbook 2013

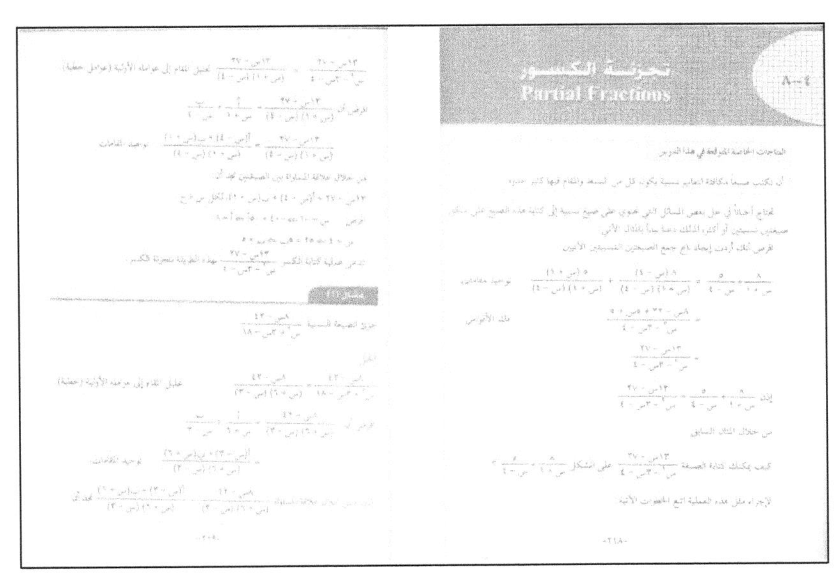
تجزئة الكسور
Partial Fractions

Random pages, 11th grade Arabic Textbook 2013

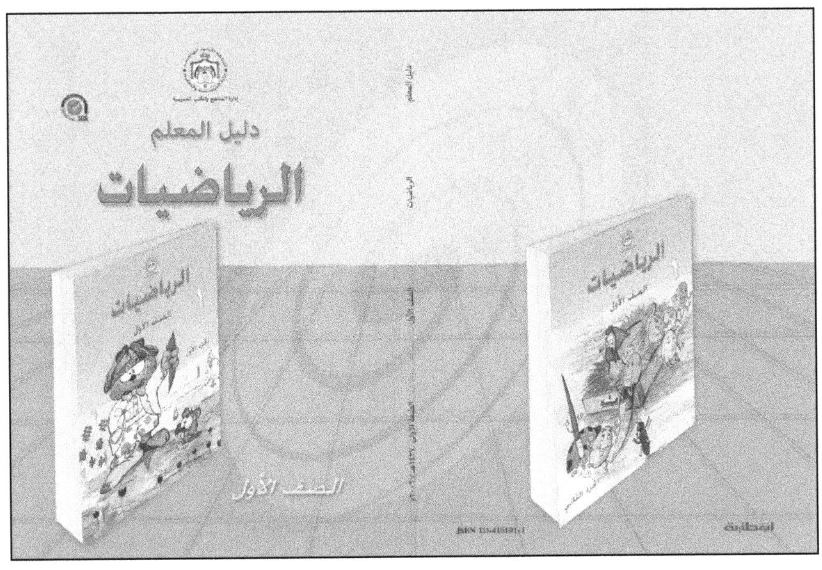

Cover, 1st grade Arabic Teacher's Guide 2013

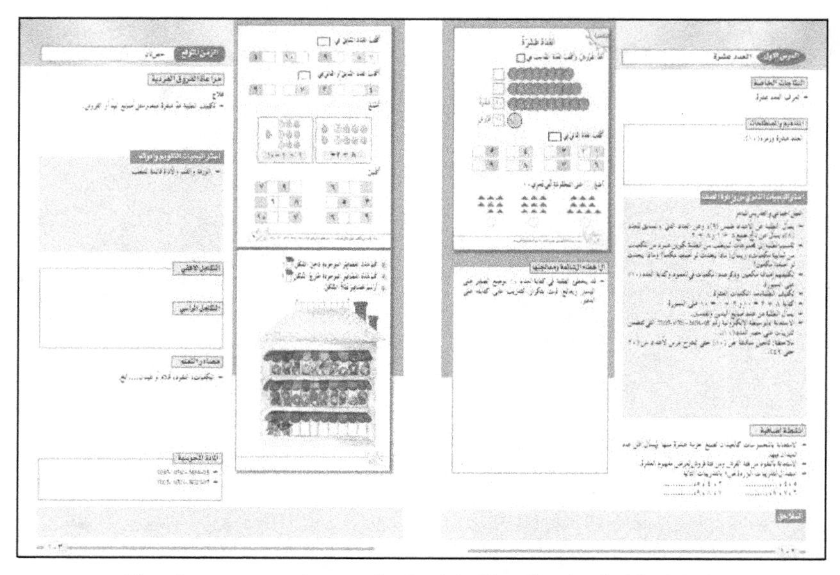

Random pages, 1st grade Arabic Teacher's Guide 2013

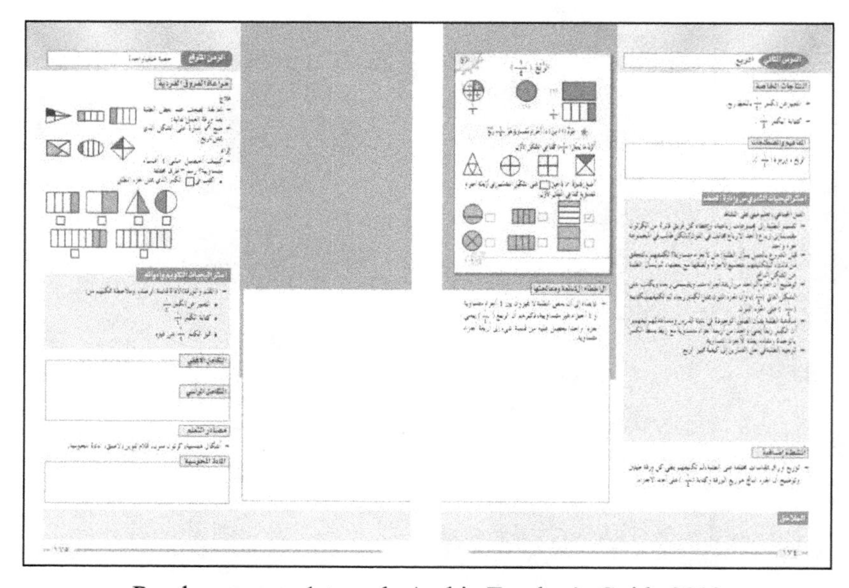

Random pages, 1st grade Arabic Teacher's Guide 2013

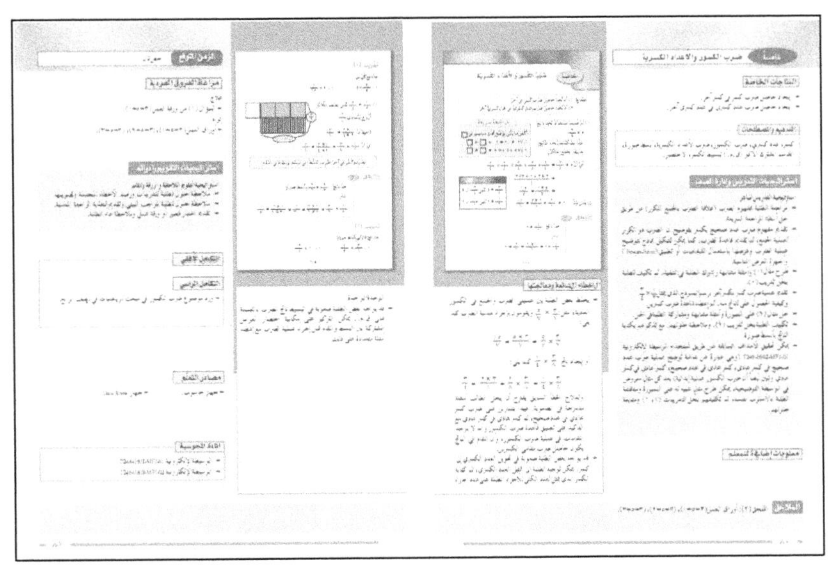

Random pages, 5th grade Arabic Teacher's Guide 2013

Random pages, 11th grade Arabic Teacher's Guide 2013

https://doi.org/10.1142/9789813146785_0007

Chapter VII

LEBANON: Mathematics and Mathematics Education in Lebanon

Hanna Nadim Haydar
Brooklyn College of the City University of New York

Introduction: Context of Lebanon

Lebanon is a small country on the eastern shore of the Mediterranean populated by approximately four million people, of which 1,005,000 are K–12 students who are taught by 96,900 teachers. (CRDP, 2014)

Lebanon's location at the crossroads of the Mediterranean and the Arab World facilitated its rich history and shaped a cultural identity of religious and ethnic diversity (McGowen, 1989). Lebanon is the most religiously diverse country in the Middle East. As of 2015 the CIA World Factbook estimates the following demographics: Muslim 54% (27% Shia Islam, 27% Sunni Islam), Christian 40.5% (includes 21% Maronite Catholic, 8% Greek Orthodox, 5% Melkite Catholic, 1% Protestant, 5.5% other Christian), Druze 5.6%, very small numbers of Jews, Baha'is, Buddhists, Hindus and Mormons.

Political life in Lebanon is characterized by sectarianism, which plays a major role in determining political and economic decisions (World Bank, 2006). The main languages spoken in Lebanon are Arabic, French, Armenian, and English with Arabic being the official language.

Figure 1. "Lebanon - Location Map (2012) - LBN - UNOCHA" by OCHA. Licensed under CC BY 3.0 via Commons.

One cannot talk about modern Lebanon without considering the unfortunate Lebanese War, which was a multifaceted civil war that lasted from 1975 to 1991 resulting in an estimated 120,000 fatalities. As of 2012, approximately 76,000 people remained displaced within Lebanon. This war also forced almost one million Lebanese to flee their country.

More recently, the Syrian crisis has had considerable impact across Lebanon. Syrians fleeing conflict continue to make up the majority of refugees in Lebanon. According to current projections, there were over 1.3 million registered Syrian refugees in Lebanon at the start of 2015. (UNHCR, 2015)

The Education System in Lebanon

Historical Background

Even though education in Lebanon has deep historical roots, many elements of the current educational system may be traced back to the early sixteenth century when the Ottoman Rule allowed French communities to open their own schools initiating a tradition of rich educational history and diversity of schools. (Zakhariya, 2008)

After World War I, Lebanon was under the French mandate until its independence in 1943. The French government could not centralize the educational system because of the complexity of the sectarian divide and the opposition of political and religious groups. This influenced the 1926 Lebanese Constitution to acknowledge freedom of education to religious confessional communities; the same was reiterated in the 1943 constitution after independence. Since that time, education in Lebanon has been characterized by the existence of parallel systems and diversified curricula. Many measures and efforts to create a more secular education were taken by trying to increase the role of the Ministry of Education following independence. These attempts, whether to centralize the educational system or to supervise private schools and enhance public schools remained inefficient, and until today education policies continue to reflect the sectarian divide. (Frayha, 2009).

This situation favored a continuation of the strong French influence on the education system that manifested in the number of educational institutions using French as the language of instruction and adopting the French educational system. On the other hand, it also allowed missionary schools to be founded by various denominations, with each defining its own system and continuing the diversity tradition. As a consequence, there are now schools in Lebanon using French, English, Arabic and even German as the languages of instruction (Abboud, 1998).

In 1968, the educational system was revised (Harik, AbuRafiq, & Bashshur, 1999) but the resulting curriculum was greatly criticized. Educational reform attempts started in the early seventies but the civil war of 1975 delayed any changes. The new, current curriculum was not passed into law until 1995. In 1995, a new reformed curriculum was introduced by the Ministry of Education and Higher Education and was approved by the House of Ministers on October 25, 1995. (Khajarian, 2011)

In 1998, another law was passed on compulsory education. Then in 2010, the Ministry of Education and Higher Education published another important document entitled Quality Education for Growth 2010–2015. This document emphasized the right to education for all and the need to ensure equal opportunities and accessibility in Lebanese education.

However, as was the case with similar laws in the past, these laws were not followed by any implementation plans and remained unfulfilled or monitored.

Educational System Structure

The Lebanese educational system is well-developed and consists of 2,789 schools. Of these, 45.6% are public, 39.3% are private tuition-based, 12.6% are private fee-subsidized schools and 2.4% are UNRWA (The United Nations Relief and Works Agency for Palestine Refugees) schools (CRDP, 2015). However, almost 66% of the 1,005,044 students in Lebanon are enrolled in private schools.

Similar to most international practices, the current structure of the education system in Lebanon divides the K–12 education into three stages: 1) Kindergarten: ages 5–6; 2) Basic education: Elementary level (includes cycle 1, grades 1–3, and cycle 2, grades 4–6) and Intermediate level (cycle 3, grades 7–9); and 3) Secondary education: cycle 4, grades 10–12.

By law, the elementary and intermediate levels are considered compulsory education, and are provided free to all students through public schools. However, mandating compulsory education is not fully enforced (Osta, 2007). Recent reports noted that Lebanon still needs to ensure that education be free, available and accessible for all, and that dropout rates be reduced from their currently very high levels (Hamdan, 2009). In a chapter that discusses Lebanon's human capital challenges, Gonzalez et al. (2008) state that even though Lebanon had made progress in raising student enrollment in the elementary and secondary sections and had been able to narrow the gender gap, the results of international assessments are not sufficient for Lebanese students to compete globally. According to the United Nation Committee on the Rights of the Child (2006), educational facilities need to be improved, especially for children with disabilities, and the government should do more to provide and improve quality public education.

All Lebanese schools are required to follow a set curricula designed by the Ministry of Education. Private schools, approximately 1,400 in all, should follow the same curricula, but may also add more courses to their curricula with approval from the Ministry of Education. However, as mentioned above, private schools manage to maintain their own parallel curricula. The public schools are financed by the Ministry of Education and Higher Education and the private schools by student tuition fees. After primary education, English or French become the mandatory language of instruction for mathematics and science for all schools (World Bank, 2006).

Mathematics Education in Basic Education: Curriculum and Assessment

The Lebanese educational reform and the new educational framework of 1995 called for new strategies in four areas: content, teaching practices, teacher and administration preparation, and assessment methods (Al-Haykaliyya, 1995). The reform addressed teaching practices that guide to the use of modern techniques and hands-on activities in classrooms (Khajarian, 2011).

The current mathematics curriculum (1995) therefore adopted learning goals which shifted away from lower-order skills that used to center merely on memorization of facts and learning of algorithms. The new goals called instead for higher-order levels of thinking, focusing on problem solving. Pedagogically, the new curriculum advocated interactive teaching methods and student-centered learning, as opposed to the traditional lecture-type teaching that dominated the old curriculum. Evaluation guidelines in the new curriculum focused on more comprehensive assessment, targeting abilities like "knowledge of facts, understanding of concepts, problem solving, and critical thinking" (Osta, 2007).

Twenty years after their inception, the new curricula were never comprehensively evaluated and their implementation was never officially assessed, researched or analyzed. Therefore, researchers and educators look for international assessments and studies to gain insights on whether the new curricula met the goals and answered the promises. One important data set comes from the Trends in International Mathematic and Science Study, TIMSS. Lebanon has been participating in the TIMSS study since 2003. Unfortunately, the results are alarming and indicate that Lebanese students are falling behind at the international level and there is a gap between the curricular goals and student achievement realities (Frayha, 2013).

One indicator of that gap is Lebanon's score for eighth Grade in TIMSS 2003.The focus of TIMSS is on problem solving and since Lebanese eighth graders of the year 2003 were actually taught from the reformed curriculum from the beginning of their elementary schooling, Osta (2007) noted that the relatively low score of 433 (lower than the international average of 466) indicated that at least the reformed curriculum goal of problem solving was not sufficiently achieved. Lebanese eighth graders were outperformed by their peers in 30 countries, and they were better problem solvers than their peers in only 14 countries. The countries' average scores ranged from 605 for Singapore to 264 for South Africa (Gonzales et al., 2004)

In her attempt to understand causes of the gap between the reform goals and student achievement, Osta (2007) conducted an assessment analysis study in which she noted how the Lebanese math curriculum reform that started by resetting of goals and objectives, revision of teaching practices, production of new national textbooks. Seven years later, the system shifted back to the old educational practices, mainly driven by the influence of the assessment system.

In Lebanon official examinations take place every year at two grade levels: the end of the intermediate cycle of study (Grade 9) for the "Brevet" certificate, which gives access to secondary school, and the end of secondary level (Grade 12) for the "Baccalaureate" certificate and graduation from pre-university education.

The results of Osta's study suggested that the old assessment culture persisted where national official examinations kept playing a high-stake role in deciding students' promotion from one cycle to another, and determining graduation from school to college. This phenomenon was affecting the new official exams and consequently every other component of the curriculum. In order to avoid the narrowing of the curriculum and the setback to the old practices there seems to be a need for reform in the assessment system that modernizes the testing policies, as well as the contents and formats of the examinations in order to align with the new curriculum directions.

Similarly, Younes (2013) analyzed the TIMSS 2007 data to examine the performance of Lebanese eighth grade students in public and private schools. Lebanon scored only 449 while the TIMSS scale average was 500. Younes found that private school students in Lebanon performed better than public school students in each mathematics content and cognitive domain in TIMSS 2007. Her findings confirmed also a positive correlation between student positive affect towards mathematics and their achievement in both public and private schools.

Commenting on Lebanon's results on TIMSS 2011 where Lebanon scored 449 for eighth grade mathematics with the international average being 500, Frayha (2013) called for more accountability measures and wondered who is responsible for the alarming state of the educational system.

Higher Education in Lebanon

Universities

The first universities in Lebanon date back to the 19th century. The American University of Beirut was established in 1866 and the University of Saint Joseph in 1877. These universities have both maintained throughout the years a high standard of education recognized all around the world.

There are currently over 40 licensed private universities in Lebanon and only one public university, in addition to public and private technical and vocational colleges leading to technical degrees. Females in Lebanon generally receive more higher education than males (Hajj & Panizza, 2002) as over 50% of all university students are female. For many Lebanese students, the prestigious institutions of higher learning are beyond reach because of the expensive tuition. Therefore, these universities are associated with education of the elite (Zakhariah, 2008).

Teacher Preparation

More than 80 percent of teachers in primary schools in Lebanon are unqualified in the main teaching subjects, including math, sciences, and languages (World Bank, 2006). Only 42 percent of public primary school teachers have a specialized degree and less than half hold a university-level degree (Alameddine & Ellis-Petersen, 2011). This places additional importance to examining different sources of teachers' professional development and preparation. There is also primarily a need for a stricter policy reform that regulates teaching qualifications, certification and retention.

Various higher education institutions including the public Lebanese university and most private universities offer various routes for teacher preparation ranging from teaching diplomas to bachelors in education to more advanced degrees (see Table 1).

In their survey of Teacher Preparation programs in Lebanon, El Mouhayar and Boujaoude (2012) noted the absence of any university level programs that prepare teachers of intermediate level (Middle School) and the lack of emphasis on field work.

Table 1.

University	BA	Masters	PhD	Teaching Diploma	Other
Al Kafaat University	√				
American University of Beirut	√	√		√	√[1]
American University of Science and Technology				√	
Global University	√			√	
Haigazian University	√	√		√	
Lebanese American University	√	√		√	
Lebanese German University	√				
Lebanese International University	√	√		√	
Lebanese University	√	√	√		
Middle East University	√	√		√	
Modern University for Business and Science	√				
Notre Dame University	√[5]	√	√	√	√[2]
Saint Joseph University	√	√	√	√	√[3]
University of Balamand	√	√	√	√	
Holy Spirit University of Kaslik	√	√	√	√	√[2] – √[4]

[1] Diploma in Special Education
[2] Teaching Certificate in Elementary and Basic Education
[3] CAPES in all subjects that are taught in the Lebanese Baccaluareate: Diplôme universitaire: Pédagogie universitaire: Démarches pédagogiques innovantes; Diplôme universitaire: Maître de stage; Diplôme universitaire: Encadrement pédagogique
[4] Diplôme universitaire en sciences de l'éducation and Formation continue en Sciences de l'éducation
[5] Specializätions in early childhood, learning disabilities, and education of the gifted.

Source: Reprinted from *Structural and Conceptual Foundations of Teacher Education Programs in Selected Universities in Lebanon*, p. 40 by El Mouhayar, R. & Boujaoude S., Recherche Pedagogique, 2012.

Preparation of Mathematics Teachers

More specifically, this section will zoom in on the preparation of mathematics teachers. To become a mathematics teacher in Lebanon, one can take one of the following available routes either at the Lebanese university or at one of the private universities:

Lebanese University

The College of Education of the Lebanese University implements the European model of higher education with the sequence License-Master-Doctorate (LMD). All programs leading to an Education License (equivalent to bachelor's degree) in elementary education require students to take 68 common credit hours in languages, general pedagogy, technology, and general culture. Students majoring in science, mathematics or language take 70 more credits of specialized pedagogy (including methods courses, content matter, and field work). In addition, students take 36 more credits of specialized pedagogy (including methods courses, subject area, and field work) in another major. Students also take six more credits considered as elective courses.

To train mathematics teachers at the secondary level, the Lebanese University has been offering the Certificat d'Aptitude Pedagogique a l'Enseignement Secondaire CAPES program (Certificate of Qualification in Education for Secondary School Teaching). To be admitted to the CAPES program students are required to hold a four-year degree in mathematics and to pass an entrance examination. After admission students complete 30 credits distributed as follows: methods courses (9 credits), practicum (4.5 credits), content (6 credits) and general education courses (10.5 credits). It is worth noting that the CAPES program for public school teachers is not offered regularly but rather based on demand for teachers in public schools. However, the same program is offered regularly for students who plan to teach in private schools (El Mouhayar & Boujaoude, 2012).

Private Universities

For teaching at the elementary level, most private universities that offer Bachelor or License in elementary education in Lebanon have general education programs. Only the Lebanese American University offers a BSc in Mathematics Education for all levels and Universite Saint Joseph offers license in teaching mathematics and science at the elementary level.

For teaching mathematics at the secondary level, teacher candidates can complete a Teaching Diploma after receiving a Bachelor of Science in mathematics at various universities. Teaching diplomas vary between 21 and 30 credits and encompass mathematics education method courses, general pedagogies and a fieldwork component.

Advanced Research, Centers and Associations

In Lebanon, there are currently various centers and initiatives of a high caliber to enhance research and dissemination of work related to advanced mathematical sciences. All of these are linked to local higher education institutions and are initiated by prominent mathematicians. It is worth noting that most of these initiatives were founded after the Lebanese War and keep functioning in spite of the shaky situation of the country. Following is the description of two centers, one association and one recent conference.

Center for Advanced Mathematical Sciences (CAMS)

The American University of Beirut established the Center for Advanced Mathematical Sciences (CAMS) in 1999, the first such center among the institutions of higher learning in the Arab world. CAMS founders noted that given the seminal historical role of the Arab Middle East in the development of mathematics and astronomy, it is only natural for the region to have such a center dedicated to advanced teaching and research. The establishment of the center was also especially timely, in view of the significant scientific talent both within the region and among its diaspora, as well as the central importance of mathematical inquiry to the region's scientific, technological, and economic development. (CAMS, 2015)

The center gained an esteemed international reputation early in its establishment. Brown University's mathematician David Mumford described it after a visit saying: "For the visiting mathematician, the most exciting part of AUB is their new Center for Advanced Mathematical Sciences (CAMS), which was started with the help of Sir Michael Atiyah and Nicola Khuri (a physicist at Rockefeller University). . . . This is a great place to visit, and I strongly recommend it" (Mumford, 2005).

By creating opportunities for top-quality research and teaching, and by encouraging academic collaboration and interdisciplinary research at AUB and in the region, CAMS has served as flagship institute within AUB's academic plan to revitalize scholarship and research in the mathematical sciences.

The objectives of CAMS are to:

- Promote original research in the Mathematical Sciences;
- Act as a focal point for collaborative research among mathematicians and scientists in the region, partly by hosting visitors for various intervals of time, and also by organizing topical meetings, workshops/seminars, and conferences in the Mathematical Sciences;
- Support pure and applied research programs in Mathematics, Computational Science, Climate Studies, Theoretical Physics, and Engineering at AUB and at other universities in Lebanon;
- Promote and contribute to Master's and PhD programs in the Mathematical Sciences at AUB;
- Foster a multidisciplinary environment encompassing disciplines that make significant use of mathematical tools;
- Identify promising new fields of Science and Engineering with strong mathematical components, and encourage their integration within CAMS and the University.

(CAMS, 2015)

The center started scientific meetings in Lebanon in various scientific disciplines like the following: First Beit Mery Workshop on Mathematical Sciences: Geometry and Physics, January 11–15, 2000; First Workshop on Dynamics and its Applications, October 21–25, 2002; First Lebanese Astrophysics Meeting '09: From Stars to Galaxies, April 14–17, 2009;

First Annual Meeting for the Lebanese Society for the Mathematical Sciences, January 29 & 30, 2010; First AUB Biomedical Engineering Winter School, February 19 & 20, 2015.

CAMS activities spanned from seminars, workshops and courses in different mathematics and mathematical sciences topics given by known prestigious international speakers from the region and all over the world.

CAMS has an international advisory currently including:

- Sir Michael Atiyah (Chair);
- Robbert Dijkgraaf (Distinguished University Professor of Mathematical Physics, University of Amsterdam, and President of the Royal Netherlands Academy of Arts and Sciences);
- Phillip A. Griffiths (Professor Emeritus, School of Mathematics, IAS, Princeton);
- Nicola N. Khuri (Professor Emeritus, Laboratory of Theoretical Physics, Rockefeller University);
- Don B. Zagier (Director, Max Planck Institute for Mathematics, and Professor at the Collège de France).

Science and Mathematics Education Center (SMEC)

Another center at the American University of Beirut, Science and Mathematics Education Center (SMEC) was established in 1969 as a recommendation by professor Milton Pella of University of Wisconsin in order to reform a traditional and stagnant science education at that time in the region. Carefully selected graduate students were trained and degreed in science and mathematics education at Wisconsin then went back and helped to establish the center.

The overall mission of the Science and Mathematics Education Center, as mentioned at its website, is four-fold:

- To conduct and support quality research on the teaching and learning of science and mathematics at the pre-school, elementary, and secondary level
- To contribute to the development of quality science and mathematics teaching and research professional

- To design and provide ongoing professional development for science and mathematics teachers in Lebanon and abroad
- To effect a positive influence on the quality and status of school science and mathematics education locally, regionally, and internationally

Since its establishment, the Center has played an innovative role in science and mathematics education in Lebanon and the region covering projects related to science and mathematics school curricula use of technology and teacher training.

The center currently engages in different activities like:

- Designing and teaching science and mathematics education course for pre-service teachers and masters level graduate students in cooperation with the AUB Department of Education.
- Designing and conducting research on teaching, learning and teacher professional development in science and mathematics
- Designing and developing instructional materials in science and mathematics for students and teachers
- Maintaining a current science and mathematics curriculum library for use by pre-service and in-service teaching professionals
- Providing outreach consultation in science and mathematics education for schools, institutions and governments regarding curriculum design, the design of instructional environment, methods of evaluation, and professional development of teachers
- Providing in-service professional development for teachers and subject-matter coordinators through special courses, workshops, institutes, conferences, or through participation in professional development initiatives by AUB or other institutions and organizations

Among the center's annual events are: Annual Science and Math Educators Conference; Annual Science, Mathematics and Technology Fair; SMEC seminars hosting international speakers of science and mathematics educators.

The Lebanese Society for the Mathematical Sciences (LSMS)

Founded in 2008 by the initiative of Dr. Fouad Al Zein of Université Paris 7, LSMS is a cross-universities association and has as a mission to gather the large community of Lebanese Mathematical Scientists in Lebanon and in the diaspora.

LSMS plans are to engage its member and the Lebanese mathematics audience in the following activities stated on their websites:

1. Strengthening the links between mathematical scientists in Lebanon and the diaspora through various activities such as, non-exhaustively: scientific meetings, summer schools, appropriate assistance to Lebanese students in order to engage them in education and research programs in Lebanon and abroad.
2. Establishing sustainable cooperations with universities and government organizations, particularly the Lebanese National Council for Scientific Research (L-NCSR) and also with the professional sectors, particularly banking and Industry.
3. Establishing sustainable relationships with similar societies and associations in countries of the region and abroad, in particular, those within the Euro-Mediterranean and American spaces, in order to renew the flow of new young mathematicians to universities in Lebanon.
4. Enhancing the publication process in high quality mathematical sciences journals, for members of the society, particularly for the young generation.
5. Publishing reviews and whenever feasible, the LSMS will also attempt publishing its own journal.
6. Organizing festivals and competitions for gifted young Lebanese children at schools and, if feasible, participation in international Olympiads for students at the secondary level.

(LSMS, 2015)

Since 2010, the society has been organizing an annual conference that rotates among the various participating Lebanese universities and shifts in focus among various mathematical fields like pure mathematics, applied mathematics and computational sciences, probability and statistics, partial differential equations, representation theory, actuarial and financial mathematics.

Lebanese International Conference on Mathematics and Applications

Another attempt to renew the role of Lebanon in the international mathematical research map, in 2015 the Lebanese University started organizing the Lebanese International Conference on Mathematics and Applications (LICMA)

Invited and selected international and Lebanese speakers presented contributions in the following mathematical fields: graph theory and applications; partial differential equations and numerical analysis; differential geometry and applications; financial mathematics and statistics; control theory.

Conclusion

As long as Lebanon's political life and governance keep suffering from the state of paralysis, corruption and divide the way we have been witnessing lately and as long as the regional crises surrounding the country remain unsolved, no official comprehensive initiatives are expected in the near future at the education level in general, let alone in the mathematics and mathematics education fields in specific. The current situation of elitism and parallel educational systems based on sectarianism will keep characterizing the education system. Gaps between public and private schools might widen even more and at the international level there is no indicator that Lebanese school mathematics will be able to translate into improved student achievement any time soon. In this sense, Lebanon's future depends on urgently needed reforms both in education and in the state institutions and public policies.

Until then, the bright side for mathematics and mathematics education in Lebanon is limited mainly to the mathematician-initiated centers, conferences and associations described above, and to the relatively large number of Lebanese mathematicians and mathematics educators at the higher education level and in the diaspora. The hope is that more of their efforts could reach and influence the K–12 mathematics teachers and students.

References

Abboud, M. (1998). Teaching mathematics in Lebanon: A post-war experience. In P. Gates & T. Cotton (Eds.). *Proceedings of the First International Mathematics and Science Education Conference* (pp. 197–206). Nottingham: Nottingham University. Centre for the Study of Mathematics Education.

Alameddine, Y., & Ellis-Petersen, H. (2011, September 8). High school teachers, quality vs. quantity. *Daily Star*. Retrieved September 16, 2015 from http://www.dailystar.com.lb/News/Local-News/2011/Sep-07/148058-high-school-teachers-quality-vs-quantity.ashx#axzz2f2HAg3KZ

Al-Haykaliyya al-Jadida li al-Ta'lim fi Lubnan (1995). [The new framework for education in Lebanon]. Beirut, Lebanon: al-Markaz at-Tarbawa li al-Buhuth, Center for Education & Research.

Center for Advanced Mathematical Sciences (2015). About CAMS. Retrieved from http://www.aub.edu.lb/cams/about_cams/Pages/about_cams.aspx

CIA. (2015). *The world factbook: Lebanon*. Washington, DC: Central Intelligence Agency. Retrieved August 22, 2015 from https://www.cia.gov/library/publications/the-world-factbook/geos/le.html

CRDP (al Markaz a-Tarbawi li-l Buhuth wa I Inrna'). (2013-2014). *Al nashra al ihsa'iyya [Statistics bulletin]*. Beirut: Ministry of Education Centre de Recherche et de Developpement Pedagogiques. Retrieved September 30, 2015, from http://www.crdp.org/en/desc-statistics/25707-2013%20-%202014

El Mouhayar, R., & BouJaoude, S. (2012). *Structural and conceptual foundations of teacher education programs in selected universities in Lebanon. Recherche Pedagogique* (Special Issue), *22*, 37–60.

Frayha, N. (2009). The negative face of the Lebanese education system. Retrieved on July 9, 2015 from https://cafethawrarevolution.wordpress.com/2009/09/14/the-negative-face-of-the-lebanese-education-system/

Frayha, N. (2013, March 18). Adaa' al-tulab al-lubnaniyin fi ikhtibarat al TIMSS [Lebanese Student Performance in TIMSS]. *Annahar Newspaper*. Beirut: Lebanon.

Gonzales, P., Guzman, J., Partelow, L., Pahlke, E., Jocelyn, L., Kastberg, D., et al. (2004). Highlights from the Trends in International Mathematics and Science Study (TIMSS) 2003 (NCES 2005–005). U.S Department of Education, National Center for Education Statistics. Washington, DC: U.S. Government Printing Office.

Gonzalez, G., Karoly, L. A., Constant, L., Salem, H., & Goldman, C. A. (2008). *Facing human capital challenges of the 21st century: Education and labor marker initiatives in Lebanon, Oman, Qatar, and the United Arab Emirates*. Santa Monica: RAND Corporation.

Hajj, M., & Panizza, U. (2002). Education, childbearing, and female labor market participation: Evidence from Lebanon. *Journal of Development and Economic Policies*, *4*(2), 43–71.

Hamdan, H. (2013). Education in Lebanon. War Child Holland. Retrieved September 15, 2015, from http://www.warchildholland.org/sites/default/files/bijlagen/node_14/31-2013/education.pdf

Harik, I. F., AbuRafiq, A., & Bashshur, M. (1999). Al-hay'ah al- lubnaniyah lil-ulum al-tarbawiya. *[The state of education in Lebanon].* Beirut, Lebanon: Lebanese Association for Educational Studies.

Khajarian, S. (2011). *A Study to Determine Differences in the Level of Perceived Preparedness in Teaching Algebra to Eighth Graders between Teachers in the United States and Teachers in Lebanon.* (Unpublished doctoral dissertation). Pepperdine University: CA.

Lebanese Society for the Mathematical Sciences (2015). History and mission. Retrieved from http://www.lsms.net/lsms/history.htm

McGowen, Afaf Sabeh (1989). "Historical Setting." In Collelo, Thomas. *Lebanon: A Country Study. Area Handbook Series (3rd ed.).* Washington, D.C.: The Division.

Mumford, D. (2005) Mathematics in the Near East: Some Personal Observations. *Notice of the AMS*, 52, 526–530.

Osta, I. (2007). Developing and Piloting a Framework for Studying the Alignment of Mathematics Examinations with the Curriculum: The Case of Lebanon. *Journal of Educational Research and Evaluation, 13*(2), 171–198.

Science and Mathematics Education Center (2015). Mission, News and Events. Retrieved from http://www.aub.edu.lb/fas/smec/Pages/index.aspx

UNHCR (2006). *Concluding Observations of the Committee on the Rights of the Child: Lebanon 3rd report.* United Nation Human Rights.

World Bank. (2006). *Lebanon quarterly update.* The World Bank Group.

Younes, R. (2013). *The Relationship of Student Dispositions and Teacher Characteristics with the Mathematics Achievement of Students in Lebanon and Six Arab Countries in TIMSS 2007* (Unpublished Doctoral dissertation). Texas A&M University: TX.

Zakharia, Z. (2008). *Languages, schooling, and the (re-)construction of identity in contemporary Lebanon.* (Unpublished doctoral dissertation). Teachers College, Columbia University: NY.

About the Author

Hanna N. Haydar is the Head of the Program in Childhood Education – Mathematics at the City University of New York-Brooklyn College. He received his BSc and Teaching Diploma from the American University of Beirut – Lebanon, and his MSc and PhD in mathematics education from Teachers College Columbia University. Dr. Haydar was previously the Curriculum Standards and Professional Development Adviser at RAND Education and the Supreme Education Council in Doha – Qatar. Dr. Haydar's research interests and publications focus on educational policy, beginning mathematics teachers, inclusion, lesson study and teachers' response to mathematical errors. He conducted many studies with the MetroMath (The Center for Mathematics in America's Cities at the City University of New York Graduate Center) examining the impact that the New York City Teaching Fellows program, an alternative teacher certification program, has on mathematics education in the city's classrooms. Dr. Haydar taught before in Lebanon, Kuwait and the United States. His international experience also includes research studies and consultancies in the Middle East, Europe and the United States. Dr. Haydar has served as a consultant to the Ministry of Education in Oman and UAE on curriculum standards development and mathematics education improvement.

https://doi.org/10.1142/9789813146785_0008

Chapter VIII

MOROCCO: Contributions to Mathematics Education from Morocco

Moulay Driss Aqil
Teachers College, Columbia University

El Hassane Babekri
Teachers College, Columbia University

Mustapha Nadmi
Teachers College, Columbia University

Introduction

Throughout history the Kingdom of Morocco has been a crossroads for many Mediterranean civilizations. The country's geographical location has caused it to be a transit point between the East and the West. Morocco has accumulated a rich educational heritage of professional formation, particularly in mathematics in which it has long traditions.

Morocco is also considered to be one of the main strongholds of Islamic civilization, its culture and its protection. The city of Fez, the scientific capital of the Kingdom of Morocco, founded in the ninth century, still stands as a witness to the sciences and scientists that Morocco introduced to human civilization. The Old Al-Qarawiyyin Mosque (Founded in 859 C.E.) was both a Mosque and a University of Sciences and Knowledge from which several scholars whose names remain immortal in the history of the Islamic world and the world's sciences and culture graduated. However, during the period when European countries began to make improvements in their educational systems, enabling them to promote their economic, industrial and cultural standard, a combination of factors came into existence. These factors made the traditional educational system in Morocco unable to keep pace with European civilization, a matter that paved the way for a colonial era. The beginning of this new era was the establishment of French schools and the adoption of a new European-style educational system.

After independence in 1956, the Moroccan nation unanimously agreed on the creation of four principles of education: Generalization, Unification, Arabitization and Moroccanization. To this day, many efforts have been made to fix the educational system and make it emulate the advanced educational standards of developed countries. In fact, this article is an attempt to shed light on what teaching mathematics looked like through these historical stages and what the most important changes during those times were, and how new reforms address these objectives.

Al-Qarawiyyin – The Oldest University

The University of Al-Qarawiyyin mosque-religious school is located in Fez, Morocco. It was founded in 859 C.E. by Fatima al-Fihria and it has been one of the leading spiritual and educational centers of the Muslim world. The university also played a leading role in cultural exchanges and transfer of knowledge between the Islamic world and Europe during medieval times.

Not only does education at Al-Qarawiyyin University concentrate on the Islamic religious and legal sciences but also offers other studies such as theology, law, philosophy, mathematics, astrology, music, chemistry, history, astronomy and languages including French and English. Some prominent students who became well-known scholars include: the historian Ibn Khaldun, the geographer Muhammad al-Idrisi, the mathematician Jacobus Golius, the Jewish philosopher Maimonides and Pope Sylvester II.

Students from all over Morocco and Islamic North and West Africa, and Muslim Central Asia attend the Qarawiyyin. Moreover, by the 14th century, there were 8,000 students studying at this university.

During the French protectorate in 1912, Al-Qarawiyyin had witnessed a decline as a religious center of learning. Despite implementing a number of structural reforms by the French administration to reform Al-Qarawiyyin between 1914 and 1947, these reforms did not modernize the contents of the curriculum or the teaching methods. Al-Qarawiyyin education was dominated by the traditional Islamic views and its scholars (the ulama) resisted the Western views imposed by the French authorities.

Al-Qarawiyyin and its associated schools (or madrasas) were integrated into the state educational system in 1947. After independence, Al-Qarawiyyin was transformed into a university under the supervision of the Ministry of Education. In 1965, with the introduction of various reforms such as modern curricula and textbooks and professionally trained teachers or faculty, Al-Qarawiyyin Schools or the madrassas were officially renamed the "University of Al-Qarawiyyin."

A Brief History of the Interest of Moroccan Scientists in Mathematics

Although it is difficult to determine the beginning of the history of mathematics in Morocco, one could say that mathematical sciences were among other subjects taught in quite early in Morocco. In his letter "The Teachers' Etiquette" published in the third century A.H., a Moroccan scholar by the name of Mohammad Bin Abdessalam Sahnoun (d. 256) emphasized the teaching of mathematics among other educational disciplines (Muhammad Haji, the Intellectual Movement in the Era of the Saadi Dynasty, Vol. 1/91). People needed it to conduct business transactions, bequests, and division of inheritances.

It is natural that mathematics was developed beyond the degree mentioned in ibn Sahnoun's letter. It soon reached the level alluded to by ibn Al Kadi during the Saadi's era. Mathematics students in Morocco at this time studied algebra, calculation by completion and balancing, including calculations with integers and roots. They used the Arabic numerals known today; they called these numerals "dust letters," (Abdullah Gannoun "Moroccan Ingenuity" pp. 289–290, Volume 1).

We will describe briefly the method of teaching general science and mathematics, with a focus on the Old Moroccan School, and also will mention some of the old pioneering mathematics scholars in Morocco.

Teaching Mathematics in Moroccan Old Schools

Many years ago, students in Morocco studied the geometry of lines, surfaces and solids. They also mastered the drawing of various geometric shapes, circles, angles, and engineering issues involving surfaces. To reinforce the science of mathematics, Moroccan mathematicians published several books in mathematics, demonstrating the exceptional efforts in this area according to each mathematician's era.

The academic content in the Moroccan Old Schools was characterized by an encyclopedic nature in addition to a variety of disciplines that combined logical and descriptive sciences. The list of sciences and general knowledge taught at these schools exceeded forty disciplines. It was formed gradually starting from the second century AH (719 AD–816 AD) with the advent of the Islamic literary and legislative sciences as well as references to sciences of earlier civilizations.

Those classifications can be regarded as a type of academic curricula which consisted of basic units, mathematics, physics, and other branches of sciences. Students could choose the discipline they want, according to a number of considerations. Lessons were conducted via records, educational systems, reading books and skimming through texts with explanation, interpretation and clarification, and discussion of its ideas, rules and concepts, through reviewing problems by study and analysis. This was done by invoking scholars' opinions in each particular problem and giving examples in order to extract a number of provisions and rules. Often, the teacher or the scholar would provide summaries of masterpieces in arts and sciences in order to pass them on to their students to preserve these arts and sciences, whether they comprehended them or not. This means that education in these schools was based on memorization and inculcation as well as on the summary of original texts, books and classifications of précis, explanations, marginal comments and records.

The teachers would outline these sources of knowledge in brief form and embedded them in a system of calculation.

Some of the Old Leading Scholars in Mathematics in Morocco

Since the 8th century, an Islamic traditional system existed in Morocco. From the 9th century to the 10th century, Islam gradually spread through North Africa to the west and south of the Sahara and along the Red Sea and the northern coast of the Indian Ocean. Before the French conquest, education played an important role in Moroccan society. Quranic and religious schools offered an Islamic traditional style of education. Islam as the main religion and Arabic as the language are two factors contributing to the identity of Morocco.

In the tenth century AD, there were a number of scientific books written by the leading mathematician of this era, Maslamah ben Ahmed ben Bassim (d. 1007). Described by his biographers as the "Astronomer and Mathematician," he introduced different terms used in algebra and standard resolutions of all six canonical equations. His works include a poem in the functions of roots and "Pollination of Ideas, Working with the Dust Letters in Mathematics," a copy of which is in the Public Library in Rabat, and all of his books are still in manuscript forms (d. 1204 AD).

In the Marinid period, there were a number of eminent scholars as well. Among them was the famous Abu Al Abbas Ahmed bin Mohammed Al Murrakuchi, known as the "Ibn Al Banna." He was a mathematician and a researcher who excelled in philosophical sciences, especially computation. Among his printed books are: *Summarizing Computation Operations*, and *Lifting the Veil on the Summary of Computation.* His books also include: *The Astrolabe Science* and *The Rule of Telling the Time by Computation.*

Among the famous mathematicians in Morocco in this era was Abu Hassan Ali bin Mohammed known as Al Qalsadi (d. 1486 AD). He is a model of the pioneers of the Muslim West in pure science, especially in mathematics. His books include: *Unraveling the Science of Computation* and *Revealing Secrets about the Dust Letters*, which is an excerpt from the former book.

Between the ninth and tenth centuries, emerged another pioneer of pure science in Morocco, Abu Abdullah Muhammad Ben Ahmed Almaknasi, known as Ibn Ghazi (d. 1513 AD).

Finally, it should be noted that in the 19th century, Morocco always kept pace with scientific movement. The translation of scientific books, from mathematics and astronomy etc. was flourishing, following the footprints of the ancestors in their interest in various knowledge of the time, including mathematics and its teaching.

Contribution of Ibn al-Banna Al Murrakuchi

In different cultures throughout history, the role of teachers in transferring mathematical concepts in an understandable manner has been crucial. One such scholar and teacher was Ibn al-Banna Al Murrakuchi, a North African figure from the 14th century. Ibn al-Banna᾿ al-Marrākushī, mathematician and astronomer, was born in Marrakech where he studied a variety of subjects, reportedly with at least 17 masters. He was one of the last innovators of the great North African mathematical tradition and at the same time, one of the initiators of a new tradition of the teaching of mathematics based on the commentaries produced by senior teachers. This tradition involved the whole Maghreb and extended itself even to Egypt and certain sub-Saharan regions. The catalog of Ibn al-Banna᾿'s works comprises about 100 titles, out of which some 50 are dedicated to mathematics and astronomy (including astrology), but the list also includes Quranic studies, theology (uṣūlaldīn), logic, law (fiqh), rhetoric, prosody, Sufism, the division of inheritances (farā᾿iḍ), weights and measures, measurement of surfaces (misāḥa), and medicine. His reputation is based mainly on his mathematical works (especially arithmetic and algebra); he has been considered the last creative mathematician in the Maghrib, meaning that he approached new problems and gave original solutions. His works were extremely popular, and inspired an enormous number of commentaries, which were still being written until the beginning of the 20th century. His works include *Talkhîs a'mal al-hisab* ,a short rhetorical presentation of operations on numbers and fractions, proportions and algebra (Concise Exposition of Arithmetic

Operations d. 1321). This famous book has been explored by researchers including George Sarton, David Eugene Smith, Renault, Marre (1864), Vernet (1952), Siouissi (1969), Francis Cajori, Aballagh M. (1994–1988), Djebbar A. (1990, 1981, 1986, 1988, 2001), Lamrabet D. (1984).

Teaching Mathematics During the Colonial Period and After Independence

The education system of Morocco has undergone several transformations over the last few centuries. Prior to the French colonization, Morocco had a long established history of Quranic based education. This traditional Moroccan education has religion and legal components (including basic mathematics). The French knew that Morocco had a long history of education and had the oldest university in the Arab world, the Quarawiyin University in Fez.

In 1912, the French Protectorate authorities established a new educational system, which greatly impacted traditional education to the extent that it led to profound changes in the economic, social, and cultural structure of the country. Significantly, under the new system only a small number of Moroccans managed to enroll in schools. Additionally, a very limited number could enroll in secondary and higher education schools. In 1952, in the region that was subject to the protection of the French, the enrollment rate in primary school was as follows: 100% of French were enrolled, 67% of Moroccan Jewish youth, and less than 10% of Moroccan Muslims (Zougarri, p.457).

During the protectorate, the French attempted to administer a reform of the Qarawiyyin University, the traditional educational institution. France's primary policy aim was to limit education and to narrow its development. The French perceived that educational freedom would be a threat to their presence in Morocco. This is one of the reasons why France did not intend for Morocco to participate in scientific and literary progress. Furthermore, under French rule, enrollment in business, science and mathematics courses was reserved exclusively for the French community, with a limited number of places for wealthy Moroccan families.

After independence in 1956, Morocco's primary goal was to establish new reforms to develop educational systems with more focus on mathematics education and science. Moreover, decisions were made to prepare teachers to qualify them to teach in Arabic. Subjects such as mathematics, natural sciences, physics and chemistry were to be taught in Arabic rather than French. In 1960, Iraq and Egypt were asked to help in the training of teachers. Egypt tried to establish a college in Rabat in which a section was reserved for the training of teachers of science and mathematics in Arabic. Iraq set up an institute for the training of teachers of history and geography.

The number of teachers increased rapidly for primary schools. However, there was still a lack of competent Moroccan teachers for secondary education where mathematics and scientific disciplines were introduced. Moreover, the Moroccan teachers were not qualified to teach scientific subjects as the terminology for these subjects was lacking from the Arabic language. Because Morocco was short of mathematics and science teachers, Moroccan authorities recruited qualified teachers from eastern European countries, such as Romania and Bulgaria.

The Importance of Mathematics Education in the Moroccan Educational System

Mathematics is part of life and has an essential social importance as it represents one of the backbones of social structure and helps in the organization and preservation of this structure. Mathematics occupies a special position in the educational system with regard to its crucial function as to the achievement of pedagogical goals and scheduling pupils' education and preparing educational policies.

Structure of the Education System

The education system in Morocco is comprised of pre-school, primary, secondary and tertiary levels. Government efforts to increase the availability of education services have led to increased access at all levels of education. Morocco's education system consists of six years of primary, three years of lower-middle/intermediate school, three years of upper secondary, and a tertiary education.

Mathematics Teaching Initiative in Pre-Primary Education

At age four children begin primary education. At this level, the goal of education is to facilitate, within two years, the physical, mental and emotional openness of the children and promote their independence and social upbringing through training of practical and technical activities such as drawing, coloring and figuration.

Mathematics Education in the Primary Cycle

Primary school education lasts six years divided into two cycles. Pupils begin primary education at pre-school, including the Quranic schools. Children who did not attend pre-school enter primary education at age six.

The first cycle of elementary school lasts two years. It aims to enforce, support, and expand the acquired knowledge during primary school to make all of the Moroccan children, once they reach the age of 8, develop a unified and coherent foundation of knowledge, to prepare them for the subsequent phases of learning. During the first two years of the first cycle the mathematics program is centered on familiarizing children with numbers and some geometric shapes and models including activities in the environment in which the children live. Through games and activities suitable for the child's development come the practice of getting used to writing, organizing, counting, measuring and identifying shapes in space and scale and their position in space and time.

Pupils who are promoted from the first cycle go on to the second cycle of elementary school. The second cycle targets, during a four-year period, mathematics, continuation of children's skill development as well as early highlights of their talents. This requires developing procedural structures for practical intelligence especially, ordering, classifying, counting, calculation and temporal and spatial orientation and operation methods.

Competencies of the Primary Cycle (Third and Fourth Year)

The most important skills to be developed in the third and fourth years of the first cycle include:

1. Time-Space Positioning;
2. Positioning according to the other and to the community institutions (family, school, community, etc.) and to adapt to them, and to the environment in general;
3. Acquiring thinking methodology and development of the mental progressive stages;
4. Acquiring the methodology of work in the classroom and outside;
5. Acquiring the methodology to organize oneself and his/her affairs and time-management of one's self-training and personal projects;
6. Mastery of the techniques of analysis, assessment, calibration and measurement;
7. Computation of simple situations that require simple mathematics;
8. Knowledge of some properties of shapes and models;
9. Sensitization to the concept of measurement through multiple handlings;
10. Mastery of other means of expression;
11. Acquisition of accurate observation;
12. Some preliminary practice about measurement, currency, length, mass and time;
13. Working with numbers from 0 to 999 in writing, naming, comparing and arranging;
14. Knowledge of the characteristics of figures and geometric shapes, including activities of assembling and dismantling simple shapes.

Teaching Mathematics in the Secondary Preparatory School

Preparatory education is a part of the secondary education and a transition between primary education and the preparatory cycle. In this sense, it represents a middle of the way in the learning path of the student. This stage consists of three academic years. It admits primary school students and prepares them to pursue their common core preparatory education.

Among the objectives of the mathematics program in this cycle is to organize and enhance students' learning and support and elevate it through mastery of the four operations on decimals and fractions, root numbers and square roots, and correct use of geometry and implementing units of measurement. The program also aims to give the learner some mathematical knowledge to enable him or her to use a real mathematical activity through gradual transition from mere calculation to algebraic style and from description to observation, experience, drawing conclusions; proving them and implementing proofs to search for solutions to a variety of mathematical problems and to deal with unfamiliar problems.

In order not to lose track of primary school and ensure cohesion and continuity, a mathematics program has been built in the three years of the secondary cycle on three main points: numerical activities—geometric activities—organizing information and numerical functions, emphasizing the fact that proportionality is a key subject in all three components and a good way to solve problems. As for terminology and symbols, they progress gradually bearing in mind the students' learning in primary school in order to guarantee unification.

Teacher Training

First cycle of Basic Education

Primary school teachers are trained at *centres de formation des instituteurs* (CFI). A *baccalauréat* is required for admission to two-year teacher-training programs and students must also pass an entrance examination. Holders of the DEUG (*Diplôme d'études universitaires générales*, an associate degree) or its equivalent can enter the second year of studies, which is mainly focused on pedagogical training. Graduates are awarded the *Diplômed'Instituteur*.

Second Cycle of Basic Education

Lower secondary teachers are trained in one of two different programs at *centres pédagogiques régionaux*. The first is a two-year program open to *baccalauréat* holders who have passed an entrance examination. Training is offered in subject specialties as well as in theoretical and practical areas. This program is only offered in subject areas where there is a particular manpower shortage; most notably, this is the case in mathematics and French. One-year pedagogical training is also offered to DEUG (or equivalent) holders who have passed an entrance examination. Graduates are awarded a *Diplôme de Professeur de Premier Cycle*.

Secondary Education

Secondary school teachers in the general education stream are taught at *écoles normales supérieurs* (ENS, higher teacher training schools); technical school teachers are trained at *écoles normales supérieurs de l'enseignement technique* (ENSET). The Faculty of Education at Mohammed V University also trains secondary school teachers. Programs can be one, two or four years in length depending on the student's qualifications upon entry into the program. Four-year programs are open to holders of the "*baccalauréat*" who have passed an entrance examination. Two-year programs are open to graduates of the DEUG (or equivalent) and to teachers of the second cycle of basic education who have sufficient work experience and have passed an entrance examination. One-year pedagogical programs are available through the Faculty of Education to graduates of the CPGE or the *license*. Graduates from all programs are awarded a *Diplôme de Professeur de Deuxième Cycle*. Technical secondary teachers are trained exclusively in four-year programs, which require an entrance examination for admission.

Mathematics Education at Teacher's Colleges and Teachers' Higher Education

Mathematics teachers play an important role in clarifying the mathematical concepts, and in the shaping of pupils' cognitive experiences, and training them to acquire basic math skills as well as in the design of experiences that motivate them to learn mathematics.

Since the '70s of the last century, Morocco has been tirelessly seeking to improve academic achievement, and qualify graduates to keep pace with progress in various aspects of human activity; however, this quest would not have occurred had it not been for the recognition of the need for good preparation for teachers and building of a new social image that addresses the responsibility placed upon each teacher.

The National Covenant for Education and Training has stated in its thirteenth "pillar" the following articles—134, 135, 136:

- Providing teachers, educational supervisors, mentors, and administrators with a solid formation before they take on their duties. This is in accordance with objectives, periods of time, formation system and training to be determined on a regular basis in light of the educational developments and pedagogical calendar.
- Training centers, enrollment standards and graduation standards are discussed.
- Reinforcing basic training and organization of continuous training sessions to make teachers capable of handling the learning requirements as well as the pedagogical and communicative skills required.
- The Education and training of professionals, regardless of their different functions or the levels in which they are engaged, benefit from two types of continuous training and rehabilitation: A thirty-hour short annual session to upgrade and enhance teachers' skills and in-depth rehabilitation sessions held at least once every three years. Table 1 lists weekly sessions for the study of mathematics at Math Teachers' Formation Centers: Years one and two.

Table 1.

Subjects	1st Year			2nd Year		
	Theory Courses	Guided Works	Applied Works	Theory Courses	Guided Works	Applied Works
Algebra and Geometry	2 h	4 h			2 h	4 h
Mathematics Analysis	3 h	6 h			3 h	6 h
Physics	3 h	3 h	2 h	2 h	2 h	2 h
Arabic, Terminology and Islamic Culture	3 h			3h		
French	2 h			2 h		
Probability				1 h	2 h	
English				1.30 h		

Source: Ministry of National Education

The Level of Academic Achievement in Mathematics

The level of academic achievement in mathematics differs from one level to another, ranging from a minimum of 25% in the second year of the preparatory school and 44% maximum achievement in the sixth year of elementary. Also, the proportions of the goals of the program for primary school fluctuate between 34% in the fourth level and 44% on the sixth level, while this ratio does not exceed the quarter in the secondary preparatory school during the second year and 29% in the third year.

It seems clear that primary students' achievement in mathematics is much better than the level of achievement of their colleagues in the secondary preparatory level, where the difference has reached 11points between the Fourth level primary and the second level preparative, and 51 points between the sixth level primary and the third level preparatory.

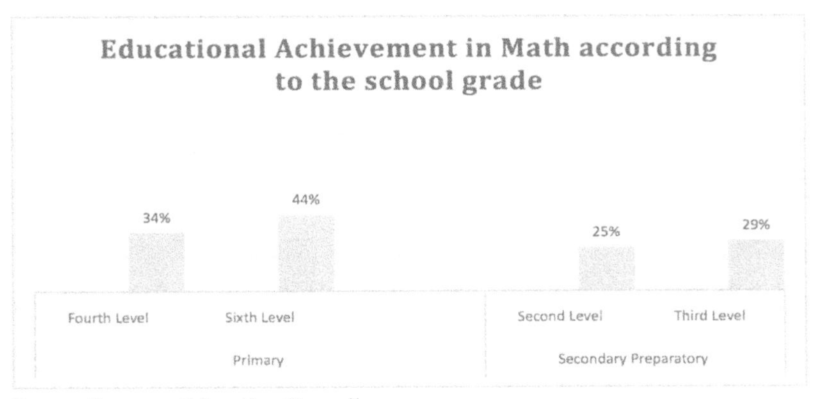

Source: Supreme Education Council

Academic Achievement in Mathematics by Gender

Overall, there has been no significant difference in terms of the level of academic achievement in mathematics between males and females as the 3-point difference which was registered in the fourth level primary in favor of males. Furthermore, as the school level gets higher, the difference between the genders is significantly decreasing. International research in mathematics and science, such as the 2003 TIMSS Assessment (Trends in International Mathematics and Science Study, a series of international assessments of mathematics and science knowledge of students around the world), showed the statistically significant differences between the genders in favor of males. Furthermore, the results obtained by Moroccan students in 2007 TIMSS, showed a difference of 4 to 8 points out of 500 between the two genders in favor of the males. However, these differences are not significant statistically.

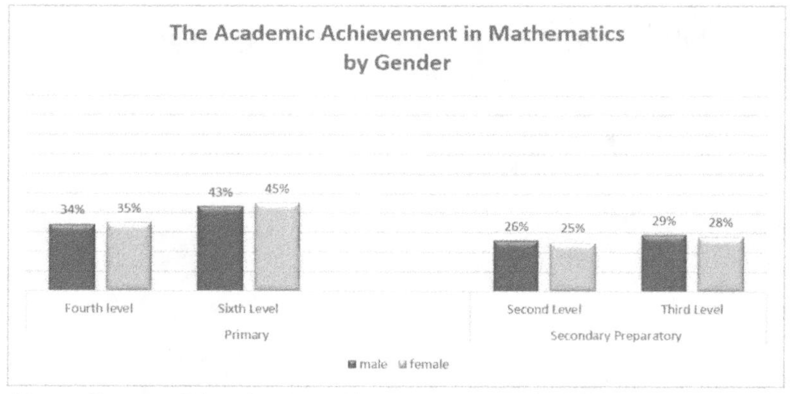

Source: Supreme Education Council

Prioritization of the Educational System by Morocco's King Mohammed VI

Since Morocco's independence in 1956, the educational system has been a major preoccupation for the King, government, and population. As successor of King Mohammed V, the Sovereign Hassan II emphasized, "Our major concern is the reform of the education system" (speech from throne, 1998). "Arabization" of the education system, the gradual adoption of Arabic language and/or incorporation of Arab culture and identity, was among one of the proposed reforms. The development of Morocco also cannot be achieved without effective knowledge of science and mathematics. For this reason, the education systems of Morocco put great importance on the study of mathematics education, science, and technology.

King Hassan II
Reign: 1961–1999.

After his accession to the throne, his majesty King Mohammed VI stated "The question of education is at the top of our present concerns because of the magnitude of its importance, its impact on the formation of generations, preparing them to take an active role in life, building the nation" (1999). In other words, the new King Mohammed VI has declared his concern for reform of Moroccan education that is strategic for the progress of society.

To meet these objectives, the King addressed the period between the years 1999–2009 as the "Education Decade" and decided to launch the National Charter for Education and Training.

King and Sultan Mohammed V. Reign: 14 August 1957–26 February 1961

As mentioned before, one of his Majesty King Mohammed VI preoccupations is to reform Morocco's education system and to ensure restoration of confidence in the school. Delivered on the occasion of the 16th throne day, the Sovereign said that "The future of the nation depends on the quality of education we provide for our children." Various policy recommendations call for changes in the education system with more focus on mathematics, changes that reflect concerns about the rigor, relevance, reasoning and sense making, and habits of minds engendered in mathematics education.

Mohammed VI. Current king of Morocco.

In addition, the Sovereign pointed out that Moroccans want an education system that is "based on open-mindedness, critical analysis and foreign language acquisition which will enable their children to access the job market and start their professional life." However, extensive and appropriate research of curriculum, teacher training, assessment and policy-making are critical in shaping the future of mathematics education, science, and technology.

The effective integration of ICT (Information and Communications Technology) into education and teaching mathematics fundamentally transforms mathematical activity. It improves student achievement on mathematics tests. For this reason, King Mohammed VI stressed that "the education reform must be aimed primarily at enabling students to acquire knowledge, skills and national and foreign languages, particularly in scientific and technical curricula, in order to be active members of society."

In a speech on the occasion of August 20th, 2015 which marked the 60th anniversary of the Revolution of the King and the People, King Mohammed VI declared that the Moroccan education system has faced challenges since its independence in 1956. The King described a new reform as one of fairness and equal opportunity, a quality education for all that will encourage individual and social progress. This educational system will combat poverty and isolation. This Strategic Vision for the Reform of the Moroccan education sector is set to take place between 2015–2030.

Conclusion

The history of educational reform in Morocco has known several ambitious reforms, producing achievements through which much has been realized. Most children now benefit from primary education and many of them have opportunities to continue their education. Great efforts have been made to establish gender equality in education by making education available to the entire population.

A clear progress towards gender equality has been achieved. The gap between boys in urban areas and girls in rural areas in primary school was narrowed by 3.5% by the year 2012. A series of programs have been launched to increase the rate of school attendance, improve the quality of education, and reform the way the education sector has been administered.

Morocco has made a major effort to improve the teaching of mathematics over the years, especially by improving the curriculum and establishing regional training centers for teachers. Moroccan leaders believe strongly that the teaching of mathematics is among the main pillars that prepare the individual to think, create and show his or her abilities while confronting problems.

The issues affecting the education today, found in this chapter, are focused on to the nature of curriculum, the nature of instruction, new forms of assessment and beliefs about mathematics. Various policy recommendations call for changes in high school mathematics that reflect concerns about the rigor, relevance, reasoning and sense-making in high school mathematics.

A historical perspective has allowed this chapter to highlight significant events in mathematics education in Morocco. Mathematics is part of Moroccan life. It has an essential social importance as it represents one of the backbones of social structure and helps in the organization and preservation of this structure. It is a highly valued subject at all levels of any educational system. Morocco has taken steps to undo and liberate mathematics and its education from the hesitance instilled by decades of the French protectorate. The dual system of schooling, imposed for decades by the French protectorate, at times continues to dictate student success and hinder performance.

Morocco has made a major effort to improve the teaching of mathematics over the years, especially by improving the curriculum and establishing regional training centers for teachers. National teacher preparation and education programs such as Genie, ICT and others are thriving, but challenges and important questions remain for research to explore. Mechanisms to support mathematics teachers in the classroom need to be promoted through systematic research. His Majesty King Mohamed VI's Strategic Vision for the Reform of the Moroccan School (2015–2030) is on target and has laid a clear landscape for the future horizons of Moroccan education. The future of mathematics education in Morocco is bright and promising for the next generation.

References

a b "Qarawiyyin". Encyclopedia Britannica. Retrieved 8 December 2011.

Aballāgh, Muḥammad (1994). Rafʿ al-hijābʿanwujūhaʿmāl al-ḥisāb li-Ibn al-Bannaʾ al-Marrākushī. Taqdīmwa-dirāsawataḥqīq.Fez.

AbdellahGuenoun, the Moroccan Ingenuity (part 1).

Abdullah Gannoun "Moroccan Ingenuity" pp. 289–290, Volume 1.

Abdel Hamid Mohiuddin (Morocco Pioneers in Pure Sciences -1- Right Call Magazine

AHMED ZOUGGARI "Educational System under French and Spanish Occupation".

A. Djebbar and MuḥammadAballāgh (2001). Ḥayātwa-muʾallafāt Ibn al-Bannaʾ al-Murrākushī[sic] maʿanuṣūṣghayrmanshūra. Rabat. (This biobibliographical survey includes a very complete list of editions and secondary literature. It updates the standard papers of H. P. J. Rénaud [1937 and 1938].).

A. Djebbar: "Mathematics in the Medieval Maghrib," AMUCHMA-Newsletter, No. 15; UniversidadePedagógica (UP), Maputo (Mozambique), 15.9.1995. Republished 30

June 2008 on www.MuslimHeritage.com as *Mathematics in the Medieval Maghrib: General Survey on Mathematical Activities in North Africa.*

Calvo, E. (1989). "La Risālat al-ṣafīḥa al-muštaraka'alà al-šakkāziyyade Ibn al-Banna' de Marrākuš." Al-Qantara10:21–50.

http://www.csefrs.ma the Supreme Council for Education, Training and Scientific Research

http://www.unesco.org

http://www.men.gov.ma the Ministry of Education and Vocational Training of the Kingdom of Morocco (Directorate of Strategy, Statistics and Planning Department of Studies and Statistics).

http://www.csefrs.ma the Supreme Council for Education, Training and Scientific Research

http://www.enssup.gov.ma/ the Ministry of Higher Education, Scientific Research and Formation of Cadres.

Ibn Khaldoun's Introduction, p. 572.

Ibn Chakroun: Aspects of the Moroccan Culture in the Marinid Era, page 218.

Illustrated Dictionary of the Muslim World, Publisher: Marshall Cavendish, 2010 [1] p. 161.

Kabbaj, M.,Talbi, M., Drissi My, M. &Abouhanifa, S. (2009). Programme GENIE au Maroc : TICE et dévеloppement professionnel.revue.sesamath.net No. 16. Retrieved, September 25, 2009, from http://revue.sesamath.net/spip.php?article 233.

Lamrabet, D. 1994a. *Introduction à l'histoire des mathématiquesmaghrébines*. Rabat, Imprimerie al-Ma'ārif al-jadīda, 302 pp.

Maghreb ArabePresse. (2007). Morocco Intel launch USD127.7Mn ICT generalization program. Available online at http://www.map.ma/eng/sections/social/morocco_intel_launc/view. Accessed March 31, 2008.

Mohamed Hajji, the Intellectual Movement during the Saadi Era: Part 91/1.

Technologies for Education for All: Possibilities and Prospects in the Arab World. Hamid Behaj.

Teachers Etiquettes – Letter published by Hassan Hosny Abdul wahhab and issued with ample translation of Ibn Sahnoun's Letter.

The Introduction of Ibn Khaldoun, Chapter XIV in Numerical Science (p. 405).

The Guinness Book of Records, Published 1998, ISBN 0-553-57895-2, p. 242.

[UNESCO World Heritage Centre. The Medina of Fez http://whc.unesco.org/en/list/170].

About the Authors

Moulay Driss Aqil is a PhD candidate in mathematics education at TC, Columbia University, New York. He holds two Master's degrees: one in Mathematics from Queens College, City University of New York and a Master's degree in Mathematics education from TC, Columbia University. He also holds a New York State Initial Certificate, Mathematics, Grades 7–12 from Queens College. He authored several articles among them "A Thought on the Struggle for Existence from the Point of View of Mathematicians, Chapter III in the G.F. Gause's 1934 Book" and "The Anchor Effect Discussed in A Mathematician Plays the Stock Market." Among his projects "Optimization Using Markov Chain Process," NYCCT, CUNY, "The Four Color Theorem" Queens College, CUNY, and "Mandelbrot versus Julia Set," TC, Columbia University. He has been teaching Mathematics at QCC, CUNY since 2009 and NYCCT, CUNY since 2011.

El Hassane Babekri is a PhD candidate in mathematics at Columbia University, New York. He holds a Master's degree in Pure Mathematics from Lehman College, City University of New York and Bachelor of Science, Physics from Mohammed Ibn Abdullah University, Fes, Morocco. He presently works at the Research Foundation of The City University of New York and as lecturer in mathematics at Bronx Community College, CUNY.

Mustapha Nadmi is a PhD candidate in Mathematics Education program at TC, Columbia University, New York. He received two Masters of Art, one in Mathematics Education from TC, Columbia University, New York and the other in Mathematics from Queens College, City University of New York. He currently Works as mathematics lecturer at New York City College of Technology, CUNY and Borough of Manhattan Community College, CUNY. Among his projects: "Generating functions and linear recurrences," Queens College, CUNY and "Chaos Theory: Universality-the Feigenbaum Constant and Renormalization" in dynamic system, TC, Columbia University. He is a member of National Council of Teachers of Mathematics and Association of Teachers of Mathematics of New York City.

MOROCCO: Appendix

A. The al-Qarawiyyin University, library, and mosque

It was originally founded as a mosque with an associated mosque school (madrasa) which has been referred to as a university. The credit of its foundation goes to a woman called Fatima al-Fihri, the daughter of a wealthy merchant named Mohammed Al-Fihri, who along with many of their community migrated from Kairouan (from where the name of the mosque has come) in Tunisia to Fes in Morocco in the early 9th century. Fatima al-Fihri together with her sibling inherited huge wealth from their father, and Fatima had vowed to utilize her inheritance on the construction and running of a mosque for her community. The dynasties that followed continued to expand the Al-Qarawiyyin mosque until it became the largest in Africa.

Fatima al-Fihriya's original diploma (a wooden board).

Original 9th-century Quran, still in its original binding.

The interior of al-Qarawiyyin University

A dedicated reading room at al-Qarawiyyin library.

Source: Samia Errazouki/AP Images.

B. Al-Qarawiyyin University: Prominent Students

Ibn Khaldoun (Historian)

Born: May 27, 1332, Tunis, Tunisia
Died: March 19, 1406, Cairo, Egypt

Ibn Khaldun was a North African Arab historiographer and historian. He is claimed as a forerunner of the modern disciplines of sociology and demography. He is best known for his book, the Muqaddimah or Prolegomena.

Ibn Khaldūn, statue in Tunis, Tun. *Kassus*

The original "Muqadimmah," a famous 14th-century text from the North African historian Ibn Khaldun in Al-Qarawiyyin library

Mohamed al Idrisi (Geographer)

Born: 1100, Ceuta, (present-day Spain)
Died: 1165, Ceuta; Known for Tabula Rogeriana

Statue of Al-Idrisi in Ceuta

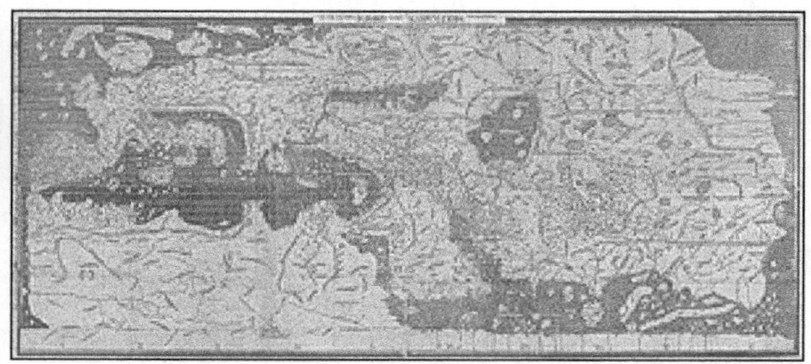

The Tabula Rogeriana, drawn by al-Idrisi for Roger II of Sicily in 1154, one of the most advanced ancient world maps.

Jacobus Golius (Mathematician)

Born: 1596, The Hague, Netherlands
Died: 1667, Leiden, Netherlands
Education: Leiden University

Jacob Golius born Jacob van Gool was an Orientalist and mathematician based at the University of Leiden in Netherlands. He is primarily remembered as an Orientalist.

Maimonides (Jewish philosopher)

Born: 1135, Córdoba, Spain
Died: December 13, 1204, Fustat, Egypt
Buried: Tomb of Maimonides, Tiberias, Israel
Education: University of Al Qarawiyyin

Moshe ben Maimon, or Mūsā bin Maymūn, acronymed Rambam, and Graecized Moses Maimonides, a preeminent medieval Sephardic Jewish philosopher and astronomer, became one of the most prolific and influential philosopher.

Pope Sylvester II

Another interesting fact is that non-Muslims were welcome to matriculate. In fact, the University's outstanding caliber attracted Gerber of Auvergne, who later became Pope Sylvester II and went on to introduce Arabic numerals and the concept of zero to medieval Europe.

https://doi.org/10.1142/9789813146785_0009

Chapter IX

OMAN: Mathematics and Mathematics Education in Oman

Nasser Said Al-Salti
Sultan Qaboos University

Hanna Nadim Haydar
Brooklyn College of the City University of New York

Introduction

Located in the southeastern part of the Arabian Peninsula, The Sultanate of Oman is an Arab-Muslim country neighboring Saudi Arabia and the United Arab Emirates to the north, Yemen to the west, the Arabian Sea to the south, and the Gulf of Oman to the East. Oman has a population of 4,224,228 with 2,359,066 Omani nationals and 1,865,162 expatriates (National Center for Statistics & Information, 2015).

Since 2011, the Sultanate of Oman consists of eleven governorates: Muscat, Dhofar, Musandam, Al Buraimi, Al Batinah North, Al Batinah South Al Dhahirah, Al Dakhliya, Al Wusta, Al Sharqia North and Al Sharqia South. Each governorate is further divided into districts (Wilayats), there are a total of 61 districts headed by a district governor (Wali).

Oman is characterized by its young population where about 50% of citizens are below the age of 24 and 30% are below age 14 (CIA, 2015).

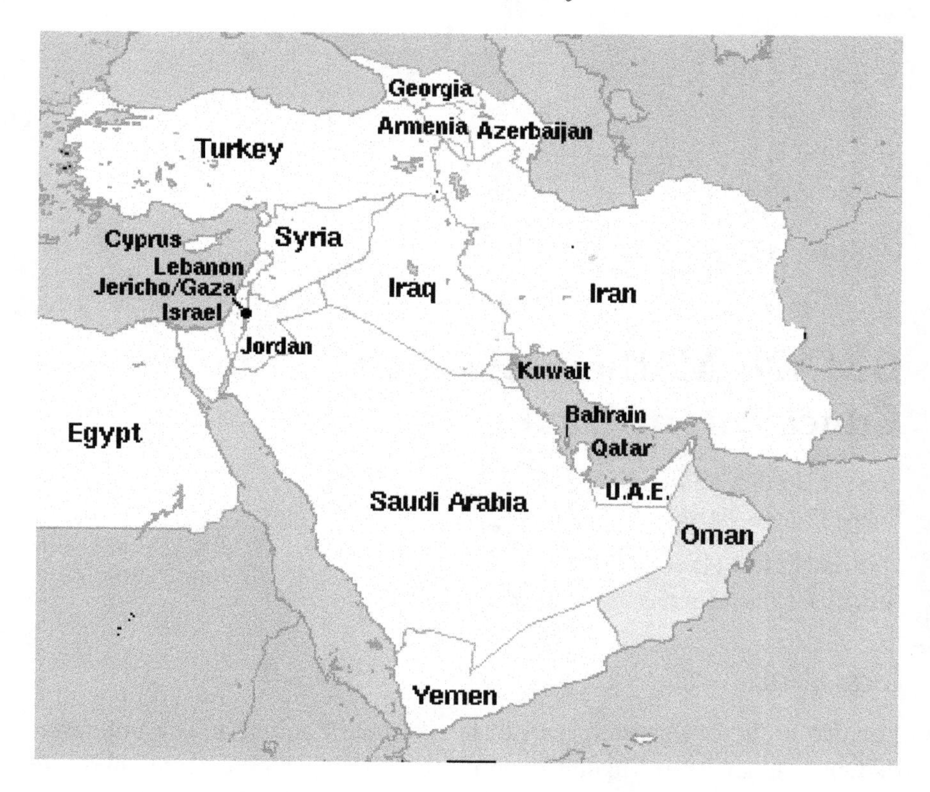

Education System in Oman

The public education system in Oman is free for citizens at all levels from first grade to the university level. The government provides free books and transportation for children in the school system. The education system consists of basic education (BE), post-basic education system or vocational education, and higher education. See Table 1.

Historical Background

The development and rise of a comprehensive public educational system in Oman goes back to 1970 after accession to power of His Majesty Sultan Qaboos bin Said. Prior to that, formal schooling was scarce and very limited. Historians traced the development of the first formal schools

Table 1. Schools in Oman.

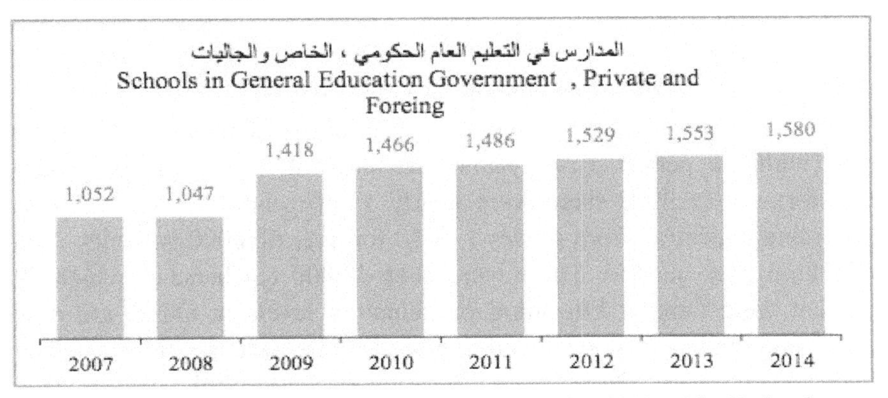

Note. Reprinted from *Monthly statistical bulletin: September 2015* p. 3 by National Center for Statistics and Information (2015)

to 1930 with the First Royal School followed by the Second Royal School in 1935, among the first schools is also the Saidya School which opened in Muscat in 1940 (Ministry of Education, 2011). Things took a major leap after Sultan Qaboos ruled the country and announced from the beginning of his ruling that disseminating education across the country and to all Omani children would be among Oman's top priorities. According to the Ministry of Education (2011), this followed with a real educational renaissance and building of schools. During the 1975–76 school year the number of schools jumped from 3 to 207, and the number of students increased from approximately 900 to 55,752 boys and girls.

According to Al Nabhani (2007), the history of education in Oman from 1970 to present day can be divided into four main stages:

1. The pre-renaissance stage: the period before 1970. There were only three formal schools and some Quranic schools.
2. The quantitative stage: the period between 1970 and 1980. The focus in education was on the rapid quantitative development of education.
3. The qualitative stage: the period between 1981 and 1995. This stage emphasized the improvement of the quality of education and the diversification of education.

4. The future stage: the period started after the "Oman 2020, The Vision Conference for Oman's Economy," many educational reforms such as the Basic Education started to address the educational requirements of the future." (p. 33)

H.E. the Minister of Education Dr. Madiha Al-Shaibani (2012) notes how within a period of 40 years "the situation has been completely transformed. In 2008 there were nearly 1,300 schools in the country, providing education from grades 1 to 12 for over 600,000 students, 48% of whom were female. There were over 43,000 teachers, of which 89 percent were Omani. Education participation levels in Oman are now equal to or above those observed in other Middle East and North African [MENA] countries" (Ministry of Education & World Bank, 2012, p. 5).

The Basic Education System

Education in Oman has undergone significant changes since the implementation of the Basic Education Reform. The Basic Education System in Oman organizes schooling years into two cycles: cycle 1 (ages 6–9) and cycle 2 (ages 9–15) followed by two years of a post-basic education cycle (equivalent to the secondary grades). This structure was first implemented during the academic year 1998–99 (Ministry of Education, 2002).

The Basic Education Reform was intended to provide:

1. Integration between theory and practice, thought and work, education and life.
2. Comprehensiveness in developing the aspects of personality.
3. The acquisition of self-learning skills in the context of lifelong education.
4. The inclusion of the values and practices necessary for mastery and excellence in learning and teaching.
5. The means to meet the needs of human development in the context of comprehensive social development. (Ministry of Education, 2001).

Some of the major features of this reform included the calls to implement student-centered activities, use a variety of formative assessment methods, revise school curricula and materials to include relevant knowledge and skills that prepare students for life and work, introduce new subject areas such as information technology and life skills, reduce class size, and equip schools with learning resource centers (Ministry of Education, 2006).

Educational Reforms

After Oman achieved impressive high-level results in terms of access to education in a relatively short time, many reform efforts and initiatives have been since directed to priorities related to education quality as measured through student learning outcomes. Her Excellency, the minister of education highlighted this emphasis on quality saying: "we believe we can successfully prepare our citizens for the great challenges facing our nation through our mission to provide them all with an Education of Quality and Excellence." (Ministry of Education/the World Bank, 2012)

In a chapter that analyzes the educational reforms in Oman, Al Balushi & Griffiths (2013) compare the educational reform efforts to tidal waves. They identify three distinct waves: the 1970s ("the borrowing stage"), the mid-1990s ("the developing stage") and from 2010 to the present day ("the collaborating phase"). They try to examine the impact of each of these waves of international influence in the context of educational reform in Oman. They explain that the recent decision to move to a new "collaborating" phase is an attempt to put in place processes that will sustain change and build long term improvement. The Omani decision-makers moved to believe that any educational reform would necessarily be relying on the creation of professional communities of teachers and schools rather than on borrowing the "baroque arsenal" (Kanan, 2013).

Mathematics Education in Basic Education

Mathematics Curriculum

Developed in alignment with the NCTM standards, the mathematics scope and sequence guides the development of curricular materials and textbooks at the Ministry of Education's Curriculum Directorate for grades 1–12.

The content of the Omani mathematics curriculum for grades 1–10 is similar to other international curricula in terms of mathematical topics and strands. The difference lies in the lack of explicit statement of process standards and mathematical practices like problem solving, representation and communication. The use of ICT in mathematics is also lacking.

At each grade level the curriculum is divided into the following mathematical strands:

- Number and Number Theory
- Number Operations
- Geometry, Trigonometry and Spatial Sense
- Measurement
- Pre-Algebra and Algebra
- Data Management and Probability

In grades 11 and 12, mathematics is still required for all students. They can select from two options and take either pure mathematics or applied mathematics. The pure mathematics track in post-basic education in Oman provides the mathematics needed for students planning to study mathematics or a related field at university. The materials of this track are again similar to other countries in terms of content but lack the emphasis on skills and mathematical practices. The applied mathematics track includes non-mathematical elements and introduces higher-level mathematics concepts that are used in business, finance and economics.

Evaluation

One important data set that allows us to examine the state of mathematical achievement of Omani students comes from TIMSS. Oman started participating in the Trends in International Mathematics and Science Study (TIMSS) since 2007 for eighth grade, and since 2011 for both fourth and eighth grades. The results were disappointing, however, and below the international average. This signaled to the decision makers the urgency of intervening to improve the mathematics and science instruction and to make the improvement of the quality of student learning the central focus of education policy.

Future Directions

In response to this gap, the Ministry of Education underwent various evaluation studies in collaboration with international organizations and is currently working on various initiatives to improve the curriculum, teacher preparation, teacher professional development and assessment. At the mathematics curriculum level, an evaluation of the current curriculum in light of international benchmarks and trends is in process. Special attention is being given to the 21st century skills and mathematical practices and on how to include them explicitly and efficiently in the curriculum.

Higher Education in Oman

Historical Background

"Let there be learning, even under the shade of trees," a famous statement of His Majesty Sultan Qaboos bin Said illustrating his commitment to provide education to all Omanis since the early 1970s when Oman's Renaissance began. The higher education history of the Sultanate of Oman has undergone rapid development. Prior to 1970, there was no formal higher education in Oman and in 1970; there were three primary schools in the country with less than 1000 students and no college or university (Al Shmeli, 2009). Between 1970s and early 1980s, higher education was achieved by offering scholarships for Omanis to study abroad. The establishment of Higher Education Institutions (HEIs) started in early eighties. The first institution was the Omani Institute of Banking (now it is called the College of Banking and Financial Studies "CBFS") which was established in 1983 and the first university, Sultan Qaboos University, was established in 1986. Now, the higher education sector consists of diverse HEIs with different curricula (Savic, 2012) offered by more than 65 HEIs, 26 out of them are private.

Administration and Management of Higher Education System

Higher education was administrated and managed by the Ministry of Education (MoE) until 1994. In 1994, a Royal Decree separated higher education from the Ministry of Education and established the Ministry of Higher Education (MoHE) to be responsible for the administration and the management of higher education system in Oman. Together with Sultan Qaboos University, they are responsible for setting up, designing and executing higher education policies in the Sultanate. At the time when MoHE was established, there were six Colleges of Education under its jurisdiction. These colleges, which were offering bachelor's degrees in education, were later converted into Colleges of Applied Sciences. MoHE also oversees private HEIs and supervises scholarships aboard.

HEIs have then taken a wide range of diversity in terms of specialization, which make them fall in terms of planning and supervision under other ministries and agencies. For example, the Ministry of Manpower supervises industrial technical colleges and vocational training, while institutes of health sciences and nursing are under the responsibility of the Ministry of Health.

Then, due to the need of unified higher education policies and for the coordination between these various HEIs, the Higher Education Council (HEC) was established in 1998 to be responsible for drawing up general higher education policies for HEIs and endeavoring to steer these policies in a way that meets the state's needs and achieves the state's cultural, social, economic and scientific goals. The HEC was later merged with the Education Council, which was established in 2012.

In the year 2000, the Oman Accreditation Council (OAC) was established to be in charge of assisting in the development of the higher education sector through institutional quality audits and institutional and program accreditation processes. In 2010, the OAC was replaced by the Oman Academic Accreditation Authority (OAAA). The OAAA was established to continue the efforts initiated by the OAC in the dissemination of a quality culture and accreditation of institutions and their programs. It is responsible for regulating the quality of higher education in Oman to ensure the maintenance of a level that meets international standards, and to encourage higher education institutions to improve their internal quality.

Early Teacher Preparation: Training Centers and The First Program

In 1970, the Sultanate had only a few options for hiring school teachers including math teachers. One of these options was to hire any Omani who has some experience in teaching and is interested to teach, regardless of his qualification. Another option was to appoint experienced teachers from other Arab countries like Egypt and Sudan (Al Rubaei, 2004). In 1972, the Ministry of Education established Training Centers for Omani teacher rehabilitation. Over 260 trainees had joined these centers, which were running the following three training programs (Al Bandari et. al, 2009):

- A one-year program for students with General Secondary School (Grade 12) Certificate.
- A two-year program for students with General Elementary School (Grade 9) Certificate.
- A three-year program for other students.

In 1975, the so-called “First Program” for Omani teacher preparation was established. It was a one-year program dedicated for preparing Omanis to teach in primary schools (Grades 1–6). It was only run for one cohort of 25 students during the school year 1975–1976 (Bahwan, 1994).

Colleges of Educations under the Ministry of Higher Education

Development Stages: From Teacher Institutes to Colleges of Applied Sciences

There were six colleges of Educations supervised by the MoHE, which were located in Ibri, Rustaq, Nizwa, Salalah, Sohar and Sur. They were offering Bachelor degrees in Education starting from the academic year 1994–1995 until they were converted to Colleges of Applied Sciences in 2007. Colleges of Education have gone through different stages. They have started as Teachers Institutes in 1976 running the following two programs (Al Bandari et. al, 2009):

- A three-year program for students who finished elementary school (Grade 9). This program first started in the school year 1977–1978 replacing a previous program for students who finished the first elementary class (Grade 7). It contained the same courses taught in secondary school plus some education courses. Graduates of this program received a certificate called "Secondary School Certificate for Teachers"
- A one-year program for General Secondary School Certificate holders. This program started in the school year 1979–1980. Graduates of this program received a certificate called "Primary School Teacher Diploma"

Teacher Institutes continued graduating primary school teachers until the end of the 1983–84 school year with a total number of 2521 teachers (Eisaan, 1995). In 1983, it was announced that Teacher Institutes were going to be replaced by Intermediate Colleges for Teachers which started effectively in the 1984–85 school year with two colleges; one in Muscat and one Salalah. By the 1992–93 school year, there were nine colleges; two in Muscat, two in Salalah and one in each of Ibri, Nizwa, Rustaq, Sohar and Sur. A student had to study four semesters in these colleges after finishing secondary school before receiving a post-secondary diploma, which qualifies him as a primary school teacher.

Starting from the 1994–95 academic year, the Intermediate Colleges for teachers were moved to Colleges of Education with the number reduced to six colleges to be under the supervision of the newly established MoHE. One of the objectives for this change was to raise the level of teaching science, mathematics and technology in the Sultanate to keep up with the scientific and technological development in the world (Eisaan, 1995). These colleges offered bachelor's degrees in education in different specialization including mathematics with program duration of normally four years.

Later in 2007, Colleges of Education were converted to Colleges of Applied Sciences (CAS). Rustaq CAS is the only College out of the six CAS colleges that still offers Bachelor of Education degrees in English language and literature. Recently, in 2015, it was decided by the Education Council to convert the Rustaq CAS back to a College of Education offering a Bachelor degree of education in different specializations including mathematics, as there are indications for a need of Omani teachers in the coming years.

Academic Programs and Math Degree Plans

There were a number of academic programs offered by the Colleges of Education during the period from 1995–2007 for the preparation of elementary and secondary school teachers (see, Al Bandari et. al, 2009 for more details). Here we will present those related to the preparation of math teachers:

A teacher preparation program with major and minor:

This program was offered during the academic years from 1995–96 to 2001–02 with the objective to prepare a teacher with the ability to teach both his major and minor subjects. A general degree plan of this program consists of 132 credits hours distributed as follows:

- Major/Minor requirements (73 credit hours)
- Practical Education (26 credit hours)
- Educational Requirements (25 credits)
- General professional courses (8 credits)

For mathematics, there were two different degree programs where mathematics was a major, namely, mathematics with a minor in computer sciences or physics, and there were two different degree plans where mathematics was minor, namely, with computer sciences or physics as a major. There are 42 credits (14 courses) in the degree plans where mathematics is a major and 18–24 credits (6–8 courses) in the degrees plans where mathematics is a minor.

A teacher preparation program with only one major:

This program was offered starting from the 2002–03 academic year. There is a degree program for mathematics with the 132 credit hours distributed as follows:

- Major requirements (80 credit hours)
- Educational Requirements (24 credits)
- Practical Education (16 credit hours)
- General professional courses (12 credits)

A program for "Raising Qualification":

This program was offered starting from the 2000–01 academic year. It was specially designed for post-secondary diploma holders graduated from the Intermediate Colleges for Teachers to obtain a Bachelor degree in education, taking into account years of experiences and focusing in major requirements. For example, for a candidate to obtain a Bachelor of Education in mathematics/physics as major/minor, he had to study 84 credits hours for a minimum of three years and for a degree in mathematics as one major only, he has to finish 93 credit hours for a minimum of three years too.

Math Education at Sultan Qaboos University (SQU)

College of Education at Sultan Qaboos University has offered a number of undergraduate programs since 1986. Postgraduate studies, on the other hand, started in the 1992–93 academic year. Regarding mathematics, it offers an undergraduate program in science and mathematics education with a specialization in mathematics since 1986, a master program in curriculum and methods of teaching mathematics since 1999 and a PhD program in curriculum and instructions in mathematics since 2012. The undergraduate math teacher preparation program has been frozen during the years 2010 and 2011 and reoffered in 2012. Undergraduate programs at SQU undergo revisions every five years. The current undergraduate degree program in mathematics education consists of 125 credit hours distributed as follows:

- Math courses (60 credit hours): 19 courses offered by the College of Science
- Educational courses (38 credits): 12 courses including practical education
- IT and Science courses (12 credit hours): three 4-credit courses offered by the College of Science
- University Requirements & Electives (12 credits): taking by all SQU students
- One English course (3 credits): English for Science

The Department of Mathematics and Statistics (DOMAS) at the College of Science at SQU offer the 19 math courses. Math degree programs offered by DOMAS will be discussed in the section "Science degree programs in mathematics." The practical education includes a full semester teaching at public schools. The yearly student intake in the undergraduate math teacher preparation program ranges from 10–15 students. Among the recent decisions made by the Education Council was to increase the student intake of the College of Education at SQU.

Math Education in Private Universities

The efforts made by the Sultanate since its Renaissance in 1970 on expanding primary to secondary education have been very successful in accommodating the growing population of young Omanis across the whole country. However, this has led to large numbers of secondary school graduates seeking higher education, which is beyond the capacity of governmental HEIs. This was made very clear by the MoHE in noting that (Al Lamki 2006):

> The greatest challenge faced by the government in human resource development in the past decade has been the widening gap between the increasing number of graduates from the secondary school system and the limited number of places available at institutions of higher education in the Sultanate.

On the other hand, there was a need to expand the base of higher education to cover different areas and multiple disciplines to meet the needs of the overall development and the labor market in the Sultanate. In response to all of this, a Royal Decree was issued in 1996 to promote the establishment of private higher education in the Sultanate and in this regard, the Omani government has encouraged the private sector to participate and play a significant role in the development of private higher education to meet the country's demand. Responding to this, a number of private HEIs were established to contribute effectively in the preparation of qualified national cadres to be able to actively participate in the development process in the Sultanate. Private HEIs, as mentioned earlier, come under the jurisdiction and supervision. A number of Royal Decrees were then issued in 1999 to formalize and regulate the establishment of private HEIs. A separate Directorate-General for private universities and colleges was then established in 2000 by the MoHE to oversee, guide and support the development of private higher education. The Private HEIs are offered a number of incentives including 50% of the paid capital of private universities. Accordingly, the number of private HEIs has increased to reach 26 institutes, which include seven universities, two university colleges and 17 higher education colleges. Private HEIs are affiliated with well recognized international universities and offer a variety of programs at different levels awarding diplomas, bachelor's and master's degrees. Regulations of offering teacher preparation programs by private universities are issued by a Ministerial Decree in 2004. Regarding mathematics, the following programs are offered:

- A Diploma in educational rehabilitation, which is a one-year program of educational courses offered to students with a Bachelor of Science degree in mathematics to be qualified for teaching mathematics in elementary and secondary schools. The availability and the student intake in this program in determined in coordination with the MoHE and MoE. It was offered several times by Dhofar University and Sohar University. It is worth mentioning that there was a HEI called "Educational Rehabilitation Institute" established in 1991 for the rehabilitation of undergraduate degree holders to be qualified as school teachers. The Institute was later closed and the program was decided to be offer in coordination with private universities. It was also recently decided by the Education Council to offer an educational rehabilitation program at SQU for different majors including mathematics starting from the 2015–16 academic year as well as at a number of private universities.

- A Bachelor of Education degree in mathematics major offered to secondary school graduates. It is currently offered by Dhofar University and University of Nizwa. The program at the University of Nizwa, for example, consists of 135 credit hours distributed as follows:
  - Math courses (47 credit hours): 15 courses offered by the Division of Mathematics and Statistics.
  - Educational courses (30 credits): 9 courses including a nine-credit practical education course.
  - University Requirements & Electives (27 credits): around 9 courses including 3 English courses and 1 course in computer skills.
  - (16 credit hours): for a minor in Computer Science, Physics or Statistics.
  - Science courses (15 credit hours): Three first courses in Computer Science, Physics and Statistics and one college elective course.

A bachelor of education degree is also offered, in coordination with MoHE and MoE, to school teachers with post-secondary diploma taking into account previous taken courses and years of experience. It was offered several times by Sohar University and University of Nizwa.

Mathematics as a Science Major at HEIs.

Science based HEIs teach a number of mathematics courses as services courses for science based programs, as part of a general foundation program and as courses for a sciences degree program in mathematics.

General Foundation Program (GFP)

Majority of students enrolling into HEIs have to undertake a form of GFP to prepare them for their further studies according to the decision made by the Higher Education Council in 2008. GFPs should be adopted all HEIs in Oman and it is a compulsory entrance qualification for Omani degree programs except those awarded by foreign HEIs. It is the responsibility of each HEI to develop a curriculum for its GFP following the standards set by the OAAA (previously OAC) in 2008, (OAC, 2008). These standards proposed that a GFP should have the following four areas of learning:

- English Language (only for English based or bilingual programs)
- Mathematics
- Computing
- General Study Skills

The aim of the Foundation Mathematics Program (FMP) is to ensure that students are equipped with the mathematical understanding and skills necessary to meet the cognitive and practical requirements of different disciplines in higher education. To meet the learning outcome standards set by the OAAA, the FMP is designed to consist of two courses. The first course is called "Basic Mathematics" and it has to be taken by all students. The second course is either "Applied Mathematics" or "Pure Mathematics" depending on the student background according to the Omani higher education system and the intended program of study. The "Applied Mathematics" course is designed for students intended to pursue their study in social sciences and humanities and the "Pure Mathematics" course is designed for students intended to pursue their study in science based programs. The content of FMP courses is of Pre-Calculus level.

Science Degree Programs in Mathematics

There are a number of science degree programs in mathematics offered by HEIs ranging from post-secondary diploma to PhD as follows:

- A Post-Secondary Diploma: Currently offered by Dhofar University as two-year program of 62 credits (27 of which are for math courses) and by the University of Nizwa as a two and half year program of 72 credits (47 of which are for math courses including a two-credit final year project)

- A Bachelor of Science: A BSc degree in mathematics is offered by Dhofar University, Sultan Qaboos University and the University of Nizwa. The BSc degree in mathematics at SQU has been offered by the Department of Mathematics and Statistics (DOMAS) since 1995. Previously, DOMAS was known as the Department of Mathematics and Computing and was offering a single degree in mathematics and computing since 1986 before the establishment of the Department of Computer Sciences in 1995. Details of degree programs are available online. Currently, DOMAS is working into offering more specialized programs in mathematics like, for example, financial mathematics, biological mathematics and industrial mathematics.
- A Master of Science: There are two master degrees offered by DOMAS; one in pure mathematics since 2001 and the second in applied mathematics since 2003.
- A PhD: The PhD program in both applied and pure mathematics is offered by DOMAS since 2009 with the first PhD student graduated in 2014.

Mathematics is also offered as a minor for the College of Science's students at SQU in which a student has to finish a minimum of 18 credits (around six mathematics courses) to be qualified for a minor in mathematics.

Conclusion

Since the 1970s, the education system in Oman has impressively transformed. Access to education reached a very high level. Omani Leadership continues to look at the education sector as a high priority. The goal is now to achieve a high-quality, efficient and relevant system. This educational emphasis is vital for the continuation and enhancement of the Omani economic growth. Calls for improvement of mathematics and science education in particular are clear within the national planning documents. Projects and initiatives are underway both at the basic and higher education levels to face the current challenges and changes. The aim is to build a culture of high standards in education and further develop the capacity of mathematics teachers and mathematics educators to improve students' mathematical achievement.

References

Al-Balushi, S. & Griffiths, D. (2013). The School Education System in the Sultanate of Oman. In G. Donn & Y. Al Manthri (Eds) *Education in the Broader Middle East Borrowing a Baroque Arsenal.* Oxford, UK: Symposium Books.

Al Bandari, M., et. al. Teacher Preparation at Colleges of Educations: A Documentary Study, Ministry of Higher Education, Oman, April, 2009.

Al Rubaei, S., Teacher Preparation and Training in the Sultanate of Oman: Achievements and Ambitions, Ministry of Higher Education, Oman, 2004.

Al Shmeli, H., Higher Education in the Sultanate of Oman: Planning in Context of Globalization, IIEP Policy Forum, 2–3 July, 2009.

Al-Lamki, S., The Development of Private Higher Education in the Sultanate of Oman: Perception and Analysis, International Journal of Private Education, 2006.

Bahwan, A. The Development of Primary School Teacher System in Oman in the View of the Experiences of England and Egypt, PhD Thesis, College of Education, Ain Shams, Eygpt, 1994.

Central Intelligence Agency (2014). The world factbook 2015. Retrieved from https://www.cia.gov/library/publications/the-world-factbook/geos/mu.html.

Eisaan, S., The Reality of Teacher Preparation, Rehabilitation and Training in the Sultanate of Oman, A Study Case, Records and Publication Committee, Ministry of Education, Oman, 1995.

Kanan, H. (2013). [Review of Education in the Broader Middle East: Borrowing a baroque arsenal by Gari Donn and Yahya Al Manthri (eds)] *Journal of Research in International Education 12(1),* 108–110.

Ministry of Education. (2002). *Basic Education in the Sultanate of Oman, the Theoretical Framework.* Muscat: Ministry of Education.

Ministry of Education. (2006). *From access to success: Education for all [EFA] in the Sultanate of Oman 1970–2005.* Muscat: Author.

Ministry of Education/ the World Bank. (2012). *Education in Oman: The drive for quality.* Muscat: Author.

National Center for Statistics and Information (2015). Monthly statistical bulletin: September 2015. Retrieved from: https://ncsi.gov.om/Elibrary/LibraryContentDoc/ben_Monthly%20Statistical%20Bulletin%20%20September%202015_b104a0cc-808d-4283-8afe-e2761c9f2b7e.pdf.

OAC, Oman Academic Standards for General Foundation Programs, Oman Accreditation Council, 2008.

Savic, M., Common Approach to QA in Diverse Higher Education Systems, Oman Quality Network Regional Conference on Quality Management & Enhancement in Higher Education, Muscat, Oman, 20–21 February, 2012.

Online Resources

1. Dhofar University: http://www.du.edu.om
2. Ministry of Education: http://www.moe.gov.om
3. Ministry of Higher Education: http://www.mohe.gov.om/
4. Ministry of Man Power: https://www.manpower.gov.om
5. National Center for Statistics and Information: http://ncsi.gov.om
6. Oman Academic Accreditation Authority: http://www.oaaa.gov.om/Default.aspx
7. Sohar University: http://www.soharuni.edu.om
8. Sultan Qaboos University: http://www.squ.edu.om/
9. The Education Council: http://www.educouncil.gov.om/en/index.html
10. University of Nizwa: http://www.unizwa.edu.om

About the Authors

Nasser Al-Salti is the Assistant Dean for Training and Community Services at the College of Science and an Assistant Professor of Mathematics at Sultan Qaboos University (SQU), Oman. He received his BSc from SQU, MSc from the University of Central Florida, USA and PhD from the University of St. Andrews, UK. Al-Salti has represented the Sultanate of Oman in the 17th General Assembly of the International Mathematical Union (IMU) held in August 2014 in Gyeongju, Korea and attended the International Congress of Mathematicians held in 2014, Seoul, Korea. He was a member of the organizing committee of the Oman Mathematics Day organized by the Oman Mathematics Committee in November 2015 at SQU. Al-Salti has also participated in the organizing committees of a number of scientific conferences in mathematics at SQU.

Hanna N. Haydar is the Head of the Program in Childhood Education – Mathematics at the City University of New York-Brooklyn College. He received his BSc and Teaching Diploma from the American University of Beirut – Lebanon, and his MSc and PhD in mathematics education from Teachers College Columbia University. Dr. Haydar was previously the Curriculum Standards and Professional Development Adviser at RAND Education and the Supreme Education Council in Doha – Qatar. Dr. Haydar's research interests and publications focus on educational policy, beginning mathematics teachers, inclusion, lesson study and teachers' response to mathematical errors. He conducted many studies with the MetroMath (The Center for Mathematics in America's Cities at the City University of New York Graduate Center) examining the impact that the New York City Teaching Fellows program, an alternative teacher certification program, has on mathematics education in the city's classrooms. Dr. Haydar taught before in Lebanon, Kuwait and the United States. His international experience also includes research studies and consultancies in the Middle East, Europe and the United States. Dr. Haydar has served as a consultant to the Ministry of Education in Oman and UAE on curriculum standards development and mathematics education improvement.

https://doi.org/10.1142/9789813146785_0010

Chapter X

SAUDI ARABIA: Mathematics Education in the Kingdom of Saudi Arabia

Dr. Mohammad Ali Alshehri
Najran University, Saudi Arabia

Introduction

Saudi Arabia is a country in the Middle East with a population of approximately 30 million people. Established in 1932 by King Abdul-Aziz Ibn Abdul Rahman Al Saud, the country covers approximately 2,240 thousand square kilometers. Arabic is the official language and Islam the official religion (Hamdan, 2005). To the west the country is bounded by the Red Sea; to the north by Iraq, Jordan, and Kuwait; to the east by Bahrain, Qatar, the United Arab Emirates, and the Sultanate of Oman; and to the south by Yemen.

The earliest education in the Arabian Peninsula coincided with emergence of Islam in the seventh century AD. Although there was no organized system of education at the time, the prophet Mohammed, peace be upon him, had urged his followers to learn the rudiments of reading, writing, and the study of religion (Dodge, 1962). Subsequently, Muslim countries were a center of scientific activity up to the thirteen century AD. Famous mathematicians such as Muhammad al-Khwarizmi began their innovations in algebra in the ninth century AD at the House of Wisdom in

Baghdad, a beacon of scientific activity during the reign of the Abbasid Caliph Ma'mun. Al-Khwarizmi was the author of *al-jabr wal-muqābala*, a famous book on algebra. To this day, significant research and many subsequent theories rely on the content of this book. As observed by George Sarton, author of *Introduction to the History of Science*, Robert Chester's translation of Khwarizmi's book into Latin in 1140 AD is, without exaggeration, the beginning of algebraic awareness in Europe. The translation was used in European universities until the sixteenth century (Aldefaa, 1981).

Until the late nineteenth century, education across what is now Saudi Arabia was traditional, restricted to reading, writing, and recitation of the Qur'an. The beginning of what one may call modern education took place at the time of Ottoman provinces of Hijaz and al-Ahsa. A few private schools in some of the larger towns began offering non-religious subjects in the 1920s, but it was not until the 1930s that modern education was sponsored by the state. A network of secondary schools was set up from 1951, and the Ministry of Education was established in 1954 (Alaklobi, 2000).

Before 1973, traditional curricula in Saudi Arabia contained arithmetic, Euclidean geometry, and algebra, virtually the same as in neighboring countries such as Egypt and Syria. Afterwards, Saudi adopted the UNESCO approach to modern mathematics and gradually applied it to schools until 1980 when, with the cooperation of university mathematicians, the Ministry of Education produced mathematics textbooks (Alqwaiz, 1984). Some examples of these textbooks are provided in the appendix to this chapter.

The result of a decision taken by the 1966 UNESCO conference was a project in Cairo in 1969 to improve the teaching of mathematics in Arab countries. The first action undertaken by UNESCO was to consider its development. Following the project, the Arab Organization for Education, Culture and Science undertook a pilot project to enhance mathematics teaching in Arab countries, beginning at intermediate and high school level as an extension of the UNESCO project, with amendments to make it appropriate to the context (Almsadiq, 1982). Next, the Arab Bureau of Education for the Gulf States saw the benefit of these initiatives and initiated new projects for the Gulf states (six Arabic countries), including one in the Kingdom of Saudi Arabia.

The Beginnings of Formal Education in Saudi Arabia

Formal education in Saudi Arabia began in 1924 with the establishment of the Directorate of Knowledge to oversee education at various stages and levels. Until then, there had been a traditional system across the Arabian Peninsula called *kuttab,* where students learnt to read and write Arabic, and to memorize the *Qur'an.* At first, these schools were open to boys only, but later they were open to girls at the lower levels. Generally, they were located either in the mosque or at the house of the teacher, who was usually the mosque's *Imam. Kuttab* schools were located throughout the Arabian Peninsula apart from the western and eastern parts, which were ruled by the Ottoman Empire during the nineteenth century. In addition to Arabic and the *Qur'an,* the curriculum sometimes included foreign languages and simple mathematics (Al Salloom, 1996; Metz, 1993).

In 1928, to meet the increasing need for teachers in schools, the Saudi Directorate of Knowledge established the first Saudi Teacher Institution. King Abdul-Aziz decided in 1935 to send Saudi students to study in Egypt and other countries, and by 1951 some 169 students were studying in Egypt to become Saudi teachers. Thenceforth, a number were sent to Lebanon, Germany, and Switzerland, and a small group of 19 to the United States (Shalaby, 1987; Wiseman, Sadaawi & Alromi, 2008).

New Era of Education in Saudi Arabia

A new era in the development of modern education began on December 24, 1953 with the establishment of the Ministry of Knowledge (replacing the Directorate of Knowledge). Its brief was to carry out the functions of planning and supervising education in Saudi Arabia, and, until the establishment of the Ministry of Higher Education in 1975 this included higher education. By 2004, the name of the ministry was changed to the Saudi Ministry of Education (Alromi & Alswidani, 2013).

Along with other members of the Arab League, in 1958 Saudi Arabia agreed upon a uniform educational system that provided six years of elementary and three years each of intermediate and secondary schooling, with a separate higher education program (Saudi Arabian Cultural Mission to the USA, 2006). The reasons for changing the curricula for mathematics in particular were accelerated growth in the different branches of mathematics, and the emergence of innovative ideas and topics applicable to all branches of knowledge. Therefore, it was necessary to implement the changes throughout the different grades. Another reason for this change in the Arab world was to avoid widening the gap between Arab and developed countries (AbuZainah, 1973), since traditional mathematics had become powerless to meet the needs of contemporary society and could no longer keep pace with scientific and technological progress.

The Educational System in Saudi Arabia

Until 1943 the educational ladder in Saudi Arabia consisted of three preparatory years, four elementary years, four secondary years, and one year of guided study. This was then amended to six primary years, three elementary years, and three years at high school (Alromi & Alswidani, 2013).

There are now five levels of education in Saudi Arabia:

1. Pre-school (nursery and kindergarten)
2. Elementary – six grades for ages 6–12 of both sexes (separately)
3. Intermediate – three grades for ages 12–15 of both sexes (separately)

4. Secondary – three grades for ages 15–18. This includes general education for both sexes (separately), and vocational (technical, commercial, and agricultural) and religious education for males only
5. Post-secondary and university.

Teacher Preparation in Saudi Arabia

The founding of the Saudi Scientific Institute in Makkah in 1924 was the first move in preparing teachers to teach in elementary schools. Ten years later, in 1934, a subdivision at the Institute created a center for the preparation of secondary school teachers. Expansion took place into the cities Onaizah, and Madinah, and in 1953 the name of these organizations was changed to "Preparing Teachers Institute." The government began to establish higher education institutions and in 1949 the first College of Sharia and Islamic Studies opened in Makkah, followed by a Teachers College there in 1952, then in other cities (Alaklobi, 2000). After this establishment of colleges and universities all over the country, teacher training programs have become an integral part of the educational system. Over the past five decades, standards for teacher training have been rising steadily, paralleling the overall development of the country's educational system. The latest minimum requirement for teaching all education levels is a four-year bachelor's degree. The schools of education at Saudi universities and colleges provide a broad curriculum in education theory and methods, and have separate departments for mathematics, physics, biology, English and Arabic language, and Islamic studies. Every student is required to major in a specialty within one of these departments, and must combine courses in education with those providing in-depth knowledge of a particular subject (Saudi Arabian Cultural Mission to the USA, 2006).

UNESCO Project for the Development of Mathematics in Arab Countries

UNESCO's General Conference of 1966 presented a comprehensive report emphasizing the need to develop the teaching of mathematics and science in Arab countries. As a result, in cooperation with experts from

Arab countries, it developed a project for the teaching of mathematics at high school level. Next, the Arab countries formed committees to participate in the project at national level to work on its preparation and implementation. The project passed through several stages, from preparing a modern curriculum of mathematics in 1969, including the study of other countries' developmental programs, devising a framework for high schools accompanied by workshops for writing and reviewing textbooks, then workshops to train teachers. This project was initiated in some Arab countries in 1970–1971 (Alsharqawi, 1988) to help them to reconsider the mathematics curriculum. It began by gradually testing new high school textbooks, then UNESCO advised the Arab countries to begin to develop the intermediate curriculum through the Arab Organization for Education, Culture and Science. This began to extend the project, aiming at the development of teaching mathematics in all Arab countries. In 1972, there were steps to prepare teachers by reconsidering the mathematics curriculum in teachers' institutes in accordance with the updated teaching curricula, developing a comprehensive plan for the training of primary school teachers so they were able to perform their mission after the changes (AbuZainah, 1973).

Most positive aspects of the UNESCO project:

- Changed the perception of mathematics as separate branches (algebra, geometry, arithmetic, and analysis) to an integrated architecture
- Introduced new concepts and used a fresh approach
- Provided an opportunity for contact and interaction between Arab and international experts in the fields of mathematics and mathematics education, which led to enriched dialogue
- Helped to make mathematical terminology consistent among Arab countries and to facilitate the development of mathematics in the various stages of education
- Led to moves in terms of training in the field to prepare for the contemporary teaching of mathematics, and introduced the whole idea in general (Alsharqawi, 1988).

Most negative aspects of the UNESCO project:

- Most authors were university professors in mathematics; very few were drawn from those working in the field or mathematics educators
- Concentrated on preparing high school students interested in undergraduate programs, and ignored other walks of life
- Involved the formation of committees from different countries, whereby each committee compiled a chapter of the book without direct contact with any others. This led to a heterogeneity of topics within a single textbook and the use of disparate languages and symbols from one subject to another, and divides between the levels of mathematical thought
- Arab states employed the proposed textbooks before testing them
- It began with high school mathematics, however it would have been better to start at primary school level, then tackle the middle and finally high school levels
- It neglected to prepare a teacher's book of guidance on teaching. (Alsharqawi, 1988)

ALECSO – Arab League Educational, Cultural and Scientific Organization Project

The UNESCO project on the development of mathematics curricula at high school level had a significant impact in terms of creating a scientific and psychological atmosphere for the development of mathematics in the Arab world. Within five years of its start, under ALECSO some Arab countries began teaching several subjects from the project and produced new textbooks on mathematics, guided by the UNESCO project. Hence, ALECSO was a pioneer in the initial development of mathematics at intermediate level (the UNESCO project covered only high schools), and expanded to cover high school levels later on (Alsharqawi, 1988).

Most positive aspects of the ALECSO project:

- The first Arab attempt to unify the mathematics curricula of Arab countries

- Went beyond the UNESCO project to include the intermediate level as well as the high school level
- Encouraged mathematics experts to meet, exchange views, and hold seminars and symposia in the field of mathematics development. (Alsharqawi, 1988)

Most negative aspects of the ALECSO project:

- There was not enough publicity about the project from the organization, therefore there was insufficient interest by the ministries of education of the majority of Arab countries
- The organization was keen to involve authorship committees from the greatest possible number of Arab states, which led to authors failing even to meet some of them and a lack of topic correlation
- The organization did not allocate enough time for representatives of Arab countries to discuss each book individually
- No vertical construction existed between the levels of courses
- No behavioral goals were set for the teaching of mathematics before deciding on the content
- No testing before implementation; also, there was no teacher's book.

Like UNESCO, the ALECSO project did not start at primary school level, then tackle elementary and finally high school (Alsharqawi, 1988).

Among the most prominent problems that the Arab states faced during the application of UNESCO and ALECSO projects were:

- Resistance from teachers to new curricula, because they already knew the traditional curriculum and there was only limited training on the new concepts
- Resistance from parents to new curricula. This might reflect their inability to help their children in the face of their ongoing complaints about the difficulty of mathematics
- Lack of interaction between students and contemporary mathematics, perhaps due to teachers who did not change their old style of teaching, as well as over-abstraction and the heterogeneity of topics within a single textbook. (Alsharqawi, 1988)

These problems arose at the start of each project, but gradually became less marked until most disappeared. Some Arab countries (including Saudi

Arabia) reconsidered the mathematics content and made necessary adjustments, or entirely rewrote it to ensure consistency and coherence between subjects. Moreover, teacher training institutions now incorporate a contemporary curriculum in their pre-service teacher preparation programs (Alsharqawi, 1988).

Arab Bureau of Education for the Gulf States Project

The Arab Center for Educational Research for the Gulf States (six Arabic countries including Saudi Arabia) has undertaken comprehensive work on the development of mathematics across the Gulf States to take advantage of the ideas emerging from previous projects.

The most prominent features of this project are:

- A workflow in accordance with scientific and educational plans, starting with proper goals then compiling content and reviewing, then training teachers to test the material and making modifications based on the results, culminating in the textbook
- Participation of representatives of all Gulf countries
- Content without exaggeration, benefitting from previous development projects, bearing in mind the introduction of books with new approaches such as an activity style and discovery learning as essential inputs in teaching
- Employment of contemporary concepts in dealing with mathematical thought, reducing the abstraction, especially at the elementary level (Alsharqawi, 1988)
- Attention to basic skills as much as concepts
- Integration of topics within the textbook, without divisions into geometry, arithmetic, and algebra, employing mathematical ideas to serve other subjects and relating them to real life situations
- Building content for understanding and reducing memorization
- Help for teachers by means of a teacher book that includes behavioral objectives and multimedia aids, accompanied by methods of teaching as well as educational activities and performance evaluation

- Takes into account the mental level of the learners and growth stages. Considers the different levels of education by starting with sensible, then semi-sensible, then finally abstraction
- Takes into account individual difference, through a companion activity book including remedial and other enrichment activities
- Attends to the issue of problem solving and encouraging self-learning through additional readings in the activity book, under the theme subject. (Alsharqawi, 1988)

Ministry of Education Project for Development of Mathematics

The most significant action taken by the General Administration of Curriculum at the Ministry of Education was establishing the so-called "National family" to carry out curriculum development. This is a technical advisory committee for each subject whose members are university professors and subject specialists from the Ministry. The first to be established was mathematics, in 1971 (Alfaleh, 1988).

The development of mathematics curricula is coupled to that of science curricula in Saudi Arabia, in the following stages. Development of science and mathematics curricula for elementary and middle school took place in 1976, when new textbooks based on modern scientific premises came out, designed to serve the local context. From 1980 to 1982 these were written for primary and elementary schools (Alfaleh, 1988). Over the three years, training courses for primary and elementary school teachers were set up in all regions of Saudi Arabia, with an accompanying teacher's book explaining how to achieve the goals of the curriculum and how to use teaching methods and aids. Finally, feedback was collected from those who implemented the materials in the field. This feedback informed the required amendments to be undertaken and the textbooks were re-printed in 1985 (Alfaleh, 1988).

The high school curriculum used the mathematics textbooks prepared by the UNESCO expert committee that inaugurated the experiment in 1974, following many training courses on modern teaching methods for teachers. From 1975 these books were gradually replaced with texts written in Saudi Arabia, starting with high schools, until by 1982 they covered all schools. Moreover, a teacher's book of training programs on the modern curriculum was issued. This took into account that the high school curricula needed to be an extension of curricula in primary and elementary schools (Alfaleh, 1988).

Research Project for Teaching Mathematics at Elementary and Intermediate Level

Funded by the King Abdulaziz City for Science and Technology, a research project for teaching mathematics at elementary and intermediate level began as a project of the National Committee for Public Education. Work on this project took three years and the project results were released in nine chapters. The project objectives may be summarized as follows:

1. Study the current situation of each:
   - Curriculum in terms of objectives, content, learning experiences, teaching methods, methods of evaluation
   - Teacher in terms of preparing, responsibilities, development and professional growth
   - Equipment helping to teach mathematics
2. Recognize the experience and expertise of developed countries in the field of mathematics education for elementary and intermediate schools, and compare it with that in Saudi Arabia
3. Develop proposals and recommendations based on the results of this study, in line with individual and society needs

At the end of the project report is a full chapter of recommendations relating to the curriculum, and teacher preparation and responsibilities, growth and development, as well as the school environment (Sahab, 2005).

Mathematics and Sciences Curriculum Development Project in Saudi Arabia

In line with the needs of global curricula development and teachers' professional development, especially in the field of mathematics and science education, there has been a worldwide wave of reforms in mathematics curriculum. Saudi Arabia, as a part of this world, adopted a mathematics and science development project, represented by the Ministry of Education as a prominent example of their strategic initiatives. This aimed at the comprehensive development of the teaching and learning of mathematics and science through curriculum development, including teaching materials, evaluation methods, e-learning development, and professional development. In order to start where the others ended, the Ministry of Education tracked the development of mathematics in various countries around the world, taking advantage of international experience to adapt in line with the vision and perspective of each country. Hence, it sought a well experienced and specialized company in this field whose products have proved effective in education improvement, namely, McGraw Hill. The material is adapted for all three stages of public education (primary—middle—secondary), and harmonized and translated into Arabic in cooperation with Obeikan, Saudi Arabia, hence, the developed mathematics curricula has been applied in the field since 2011 and is expected to end in 2016.

The project philosophy is based on ten principles:

- Student-based learning
- Excitement-based multimedia
- Learning through multiple pathways
- Learning through collaborative work
- Knowledge sharing, networking and representation by multiple methods
- Active learning, based on exploration and induction
- Development of thinking skills
- Development of decision-making skills
- Development of learners' ability through planned initiatives
- Learning through real life contexts. (Alshaya & Hamid, 2011)

The project aims to achieve the following:

- Construction of mathematics and sciences curricula with supporting educational materials, such as special mathematics and sciences textbooks, teachers' manuals, activity brochures, and scientific experimentation, transparencies, educational CDs of international standard
- Construct the latest institutions and scientific research centers to achieve standards and evaluation processes at an international level

- Take advantage of prominent specialized international expertise in the production of educational materials, and employ technical support in the implementation of mathematics and natural science curricula in public schools
- Teacher and supervisor professional development, along with continuous support from international specialized experts and appropriate training, up to international standard. The philosophy is in terms of teaching strategies, evaluation, class management, and technology integration
- Improvement in students' learning level in accordance with active learning principles, enhancing self-learning to access and build knowledge. (Alshaya & Hamid, 2011)

Universities' Role in Mathematics Education Reform

At the beginning of the 1970s, Saudi Arabia initiated massive curriculum reform with the help of its universities and a growing number of young men, educated mostly in the West and enthusiastic to develop their country (Almugwashi, 1978). In mathematics education, there has been reform over all the Arab states initiated by UNESCO's mathematics project to assist their efforts to achieve curriculum reform in secondary schools (UNESCO, 1969, pp. 4–5). In 1971, the Ministry of Education formed a National Society for the Development of Mathematical Education (NSDME). Its members consisted of mathematicians from the University of Riyadh (later, King Saud University) and mathematics specialists from the Ministry of Education. By 1972, NSDME decided to introduce modern mathematics to selected secondary schools. As mentioned above, NSDME adopted the Egyptian version. It set 1973–1974 as the date to complete the textbook for the tenth grade, continuing the remaining texts over the following two years. It produced a total of six textbooks: two for the tenth grade; two for the eleventh; and two for the twelfth (Almugwashi, 1978).

Prominent University Faculty Members Participating in the Reform

Prof. Mohammed A. Al-Gwaiz, Professor of Mathematics at King Saud University, was awarded his BS in electrical engineering by the University of Texas, Austin, USA in 1964, his MS in mathematics by Courant Institute, New York University, and his PhD in mathematics from the University of Wisconsin, Madison, USA, in 1972. He was Chair, National Society for Mathematics, Ministry of Education, and participated in authoring mathematics textbooks for intermediate and high schools. In addition, he was a member of the National Preparatory Committee for Education Policy, Ministry of Education, and Member, National Evaluation Team for Education, and currently Editor-in-Chief of the Arab Journal of Mathematical Sciences. He is the author of nine university-level textbooks on Arabic and English language.

Prof. Ali Abdullah Al-Daffa' was awarded his BSc by the Stephen F. Austin State University in 1967 in mathematics, then his MSc from East Texas State University in 1968, and his PhD by the Peabody College of Vanderbilt University in mathematics education in 1972. He has worked at King Fahd University of Petroleum and Minerals in Dhahran (Saudi Arabia) since 1972. Prof. Daffa' has twice been elected President of the Union of Arab Mathematicians and Physicists (1979–1981 and 1986–1988). He is a member of many organizations and committees, and the author of 36 books on mathematics and the history of sciences (32 in Arabic, 4 in English), and more than 250 articles in international and Saudi journals.

Dr. Salman A. Al-Salman was awarded his BSc in 1965 and his PhD in 1973 in pure mathematics, University of Birmingham, England. He is the author or co-author of 16 books on mathematics in the Arabic language. He participated in authoring five mathematics textbooks for teacher training colleges (intermediate colleges, and later teachers colleges) in Saudi Arabia, and 27 textbooks (teacher manuals) for elementary, intermediate and high schools in Saudi Arabia and the Gulf States.

Prof. Abdallah A. Al-Mugwishi, was awarded his BS in mathematics by Portland State University, Oregon, USA in 1971, his MA in mathematics education and economics by Portland State University, Oregon, USA in 1973, and his EdD in mathematics education/general curriculum from the University of Northern Colorado, USA, 1978. He was Head of the Department of Curriculum and Instruction, College of Education, King Saud University (KSU) for several years, and served as member of the National Council of Mathematics and the Ministry of Education. He was Chair of the Committee on Revision of Curriculum and Contemporary Mathematics Books for the Elementary and Intermediate Levels, Ministry of Education. He has written eight textbooks and published more than 18 research papers.

Mathematics Programs Prominent in the Reform

The Department of Mathematics at King Saud University was established in 1958. It was one of the earliest departments to be established there and, at present, is one of the largest departments of mathematics in all the Gulf States, offering BSc, MSc and PhD degrees in different areas. Since its inception, the Department has developed continuously and serves an ever-increasing number of students from the various colleges of the University. Currently, there are 77 faculty members comprising professors, associate professors, and assistant professors specializing in various mathematical areas. The Department has also 46 lecturers and teaching assistants. Apart from teaching, faculty members are actively involved in research in various fields, published in reputable international journals. Some are on editorial boards and many are reviewers of various journals of international repute.

The Department has been involved in the provision of many services sectors and various community organizations, including the development of mathematics curricula at various stages of pre-university education, and authoring and evaluating textbooks in various stages of education. AlGwaiz (1984) indicated that the Department of Mathematics at the University of Riyadh (later, King Saud University) has a prominent role

Academic Ranking of World Universities in Mathematics - 2015

Mathematics | Physics | Chemistry | Computer | Economics | Methodology | Statistics

World Rank	Institution	Country /Region	Total Score	Score on Alumni
1	Princeton University		100.0	88.2
2	Stanford University		89.4	0.0
3	Harvard University		87.2	100.0
4	University of California, Berkeley		82.6	47.1
5	Pierre and Marie Curie University - Paris 6		81.4	57.7
6	King Abdulaziz University		79.2	0.0
7	University of Oxford		72.1	40.8
8	University of California, Los Angeles		71.9	0.0
9	University of Cambridge		71.4	74.5

in the development of syllabi for secondary school mathematics, the preparation of textbooks, and teacher training.

Another example is the Department of Mathematics at King Abdulaziz University, established in 1973. This has undergone continuous improvement and achieved an international reputation by being awarded tenth place in 2014 and seventh place in 2015 in the Best Global Universities for Mathematics, published by Shanghai Jiao Tong University, also known as the Academic Ranking of World Universities (ARWU).

Finally, the Kingdom of Sandi Arabia has established 28 governmental universities, and more than ten private universities and colleges in which mathematics programs are the norm. The country continues to improve mathematics syllabi and their content, as required.

References

AbuZainah, F. (1973). Mathematics Curricula: Present and Future. Teacher Message – Jordan, vol 16, pp. 27–33.

Al Salloom, H. (1996). Education and Learning in Saudi Arabia. Riyadh: Ministry of Education, Saudi Arabia.

Alaklobi, Fahd Abdullah (2000). Education in the reign of King Abdulaziz. Scientific Journal of King Faisal University. Special issue on the occasion of a centenary of the foundation of Saudi Arabia.

Aldefaa, Ali Abdullah (1981). Introduction to the history of mathematics at the Arabs and Muslims (book in Arabic) message institution in Beirut, pp. 105–111.

Alfaleh, Naser (1988). The development of curricula and study plans in stages of public education at the Ministry of Education during the last ten years. Documentation Educational Magazine, vol. 29, pp. 69–129.

Almsadiq, A. (1982). Problems and obstacles to the development of mathematics curricula in Arab countries, Educational mission, morocco, p. 65–72.

Almugwashi, A. (1978). An analysis of the mathematics Curriculum in the public secondary schools in Sandi Arabia. Ed.D. thesis, University of Northern Colorado, Greeley, Colorado, USA.

Alqwaiz, Mohammad (1984). An overview of contemporary mathematics in Saudi Arabia. Contemporary Mathematics symposium – Riyadh, Saudi Arabia.

Alromi, A; Alswidani, Amir (2013). Glimpses of the march... Saudi education during the 90 years. The knowledge magazine 10/10/2013 [online] retrieved Aug. 12, 2015 at http://www.almarefh.net/index.php?CUV=418&Model=M.

Alsharqawi, A. (1988). Contemporary Mathematics—its Nature, and its Teaching Problems in the Elementary Schools. The teaching of mathematics and physics seminar proceeding in public education in the GCC States, pp. 361–401.

Alshaya, F.; Abdul Hamid, A. (2011). The development of mathematics and natural science curricula in Saudi Arabia Project: hopes and challenges. 15th scientific conference (scientific education: a new thought for a new reality) – Egypt, pp. 113–128.

Dodge, B. (1962). Muslim Education in Medieval Times. (Washington D.C.: The Middle East Institute, p. 1.

Hamdan, Amani (2005). Women and education in Saudi Arabia: Challenges and achievements. International Education Journal, 6(1), 42–64.

Metz, H. C. (1993). Saudi Arabia: A Country Study (5th ed.). Washington, DC: US Government Printing Office.

Sahab, Salem (2005). The final report summary of mathematics education project for primary and middle schools in Saudi Arabia. [Online] retrieved on Aug. 12, 2015. at http://www.t1t.net/download/c26.doc.

Saudi Arabian Cultural Mission to the U.S.A. (2006). Educational System in Saudi Arabia. Washington D.C.: The Saudi Arabian Culture Mission to the U.S.

Shalaby, Ali Muhammed (1987). History of Education in the Kingdom of Saudi Arabia. Dar Alqalam: Kuwait.

UNESCO (1969). School Mathematics in Arab Countries. Publication No. SC/VIS/201, Paris.

Wiseman, A.; Sadaawi, A; Alromi, N. (2008). Educational Indicators and National Development In Saudi Arabia. Paper presented at the 3rd IEA International Research Conference 18–20 September 2008 – Taipei City, Taiwan.

About the Author

Dr. Mohammad Ali Alshehri is the Dean of Scientific Research and associate professor of mathematics education at Najran University, Saudi Arabia. He holds a doctoral Degree in Teaching Mathematics at a College Level, from Teachers College, Columbia University in the city of New York, and master degree in Mathematics from Western Illinois University. Dr. Alshehri has served as Dean of Development and Quality, Dean of Graduate studies, at Najran University, Dean of College of Technology at Najran.

Dr. Alshehri has published numerous research papers in the field of Mathematics Education, including the Use of Modern Technology, and e-learning in Teaching and Learning of Mathematics, and many article in the field of developed mathematics curriculum, and has attending many scientific conferences inside Saudi Arabia and abroad like USA and China. Finally, he supervised several theses for master degree students.

SAUDI ARABIA: Appendix

A. Table of contents for Modern Mathematics for grade 10, first semester and second semester 1981

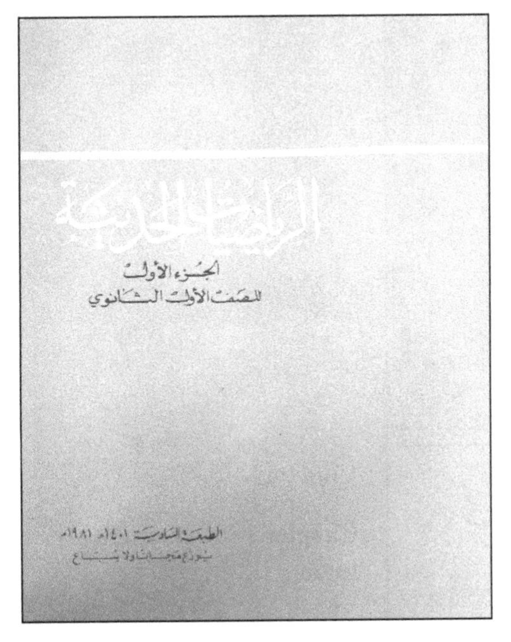

[Title] *Modern Mathematics*
Part-1 (first semester)
Grade 10

Sixth Edition, 1981

Not for sale and distributed free

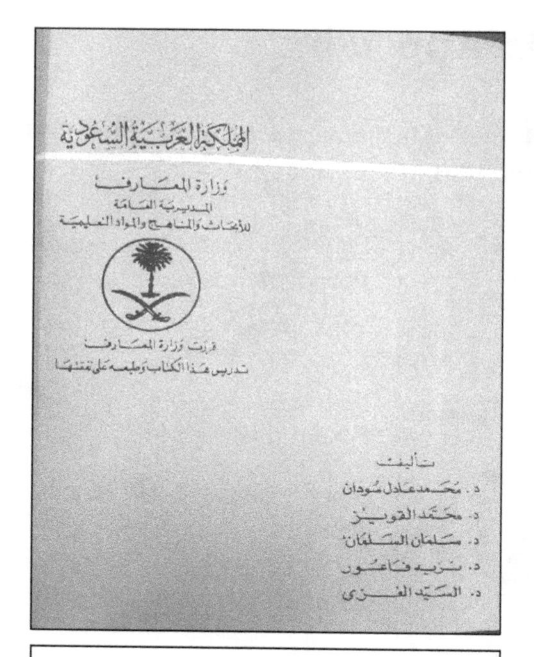

المملكة العربية السعودية

وزارة المعارف

المديرية العامة

للأبحاث والمناهج والمواد التعليمية

قررت وزارة المعارف

تدريس هذا الكتاب وطبعه على نفقتها

تأليف

د. محمد عادل سودان

د. محمد القويز

د. سلمان السلمان

د. نزيه فاعور

د. السيد الغزي

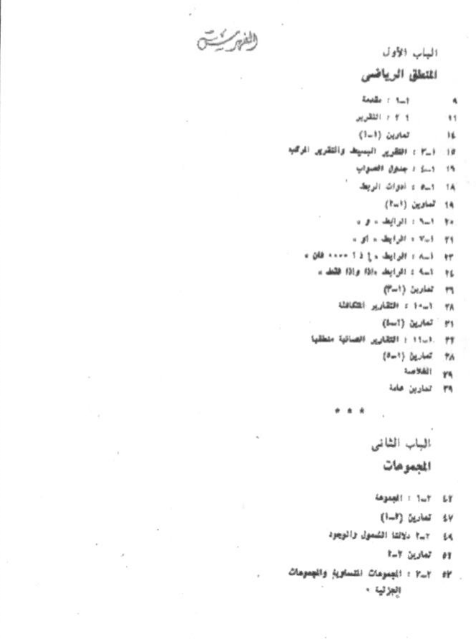

الفهرس

الباب الأول

المنطق الرياضي

* * *

الباب الثاني

المجموعات

Content:

المملكة العربية السعودية
وزارة المعارف
المديرية العامة
للأبحاث والمناهج والمواد التعليمية

الرياضيات الحديثة
الجزء الثاني
للصف الأول الثانوي

قررت وزارة المعارف تدريس هذا الكتاب وطبعه على نفقتها

الطبعة الثالثة ١٤٠١هـ/١٩٨١م
يوزع مجاناً ولا يباع

المملكة العربية السعودية

وزارة المعارف
المديرية العامة
للأبحاث والمناهج والمواد التعليمية

قررت وزارة المعارف
تدريس هذا الكتاب وطبعه على نفقتها

مراجعة:
د. سلمان السلمان
د. نزيهة فاعور
الأستاذ عبد العزيز الصالح

تحرير:
د. محمد القويز
الأستاذ محمد ضميم

تأليف:
د. السيد الغزي
د. محمد القويز

[Title] *Modern Mathematics*
Part-2 (second semester)
Grade 10

Third edition, 1981

Not for sale and distributed free

فهرس

الباب السابع
المعادلات والمتباينات

* * *

الباب الثامن
الهندسة التحليلية

* * *

الباب التاسع
التحويلات الهندسية

* * *

الباب العاشر
حساب المثلثات

★ ★ ★

Content:

B. Table of contents for Mathematics for grade 10, first semester and second semester 2014

First Semester Part 1

[Title] *Mathematics*
Grade 10 First Semester

2014 edition, the ministry of education in KSA has decided to teach this book and print it at its own expense

Distributed free and not for sale

الفهرس

1 التبرير والبرهان

2 التوازي والتعامد

Content:

الفهرس

3 المثلثات المتطابقة

4 العلاقات في المثلث

Second Semester Part 2

[Title] *Mathematics*
Grade 10 Second Semester

2014 edition, the ministry of education in KSA has decided to teach this book and print it at its own expense

Distributed free and not for sale

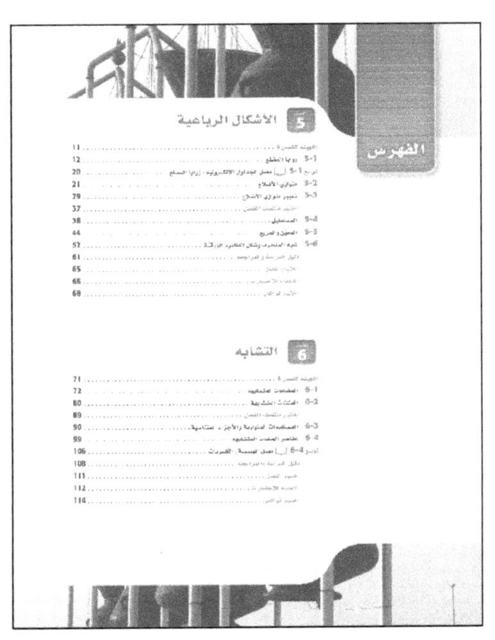
الفهرس
5 الأشكال الرباعية
6 التشابه

الفهرس
7 التحويلات الهندسية و التماثل
8 الدائرة

Content:

https://doi.org/10.1142/9789813146785_0011

Chapter XI

SENEGAL: Weaknesses and Strengths of Mathematics Education in Senegal

Serine Ndiaye
Teachers College, Columbia University

Bakary Sagna
Teachers College, Columbia University

Senegal, like many African countries, has been shaped by its colonial experiences in both geographic and cultural ways. Since its independence in 1960, the country has continued to use the French model of education. The teaching and learning of mathematics also comply with the French model. Programs and structures are modeled after the French system, which provides many substantial supports in the form of scholarships, grant programs, exchanges, modern research tools and facilities. At times, Senegalese mathematicians are invited to spend times at French institutes. One disadvantage is that the French-based curricula is considered outdated, underemphasizing modeling and interdisciplinary studies. Furthermore, the best Senegalese scholars are often tempted to remain in France draining scarce local talent. According to some faculties, depending on colonial powers reduces local incentives to think independently and solve Senegalese problems.

Like most African countries, mathematical development suffers from low numbers of qualified secondary school teachers and mathematicians

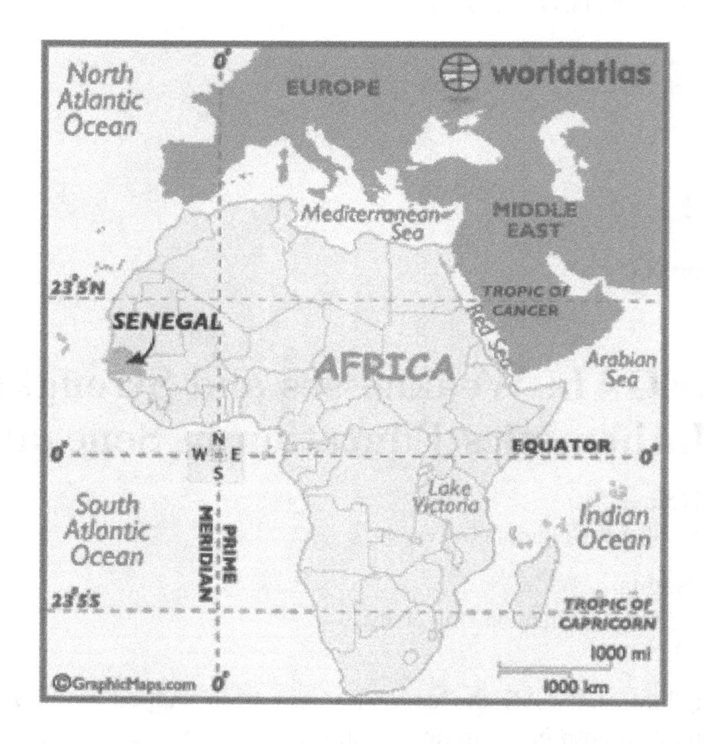

at the masters and PhD levels. Because there are few professors to train the next generation of leaders in the mathematics field, the country cannot meet the growing demand for mathematicians with advanced, up-to-date training. Talented students are not motivated to pursue a career in mathematics because of low salaries, a poor public image and a shortage of mentors engaged in exciting mathematical challenges. Most building and classrooms were built more than fifty years ago and are not adequate for today's demand. At universities, classrooms initially designated for 40–50 students are commonly crowded with at least 300 students per class. Additionally, most instruments for science courses date from the colonial days of 1960s, limiting the ability to teach the mathematics. Similarly, few mathematics students have access to even rudimentary computing facilities.

The earlier education system (1960–1975)

In Senegal the first modern school was opened in 1817, and Ecole Des Notables was established in 1841 to educate civil servants and the sons of officials. At the time, general education aimed at educating administrators to fill the increasing number of posts available, and thus attention given to sectors such as agriculture and teacher education was not significant.

Namuddu and Tapsoba (1993) argued that teacher education in Francophone countries in general have some shortcomings. For example, very few universities have faculties of education; training in education, which is based on the French model, is provided in teacher training colleges, and focuses on classroom approaches. Further training in educational action research for teachers is not conceived as a means of promotion and/or problem solving, as promotion is based on years of service. This situation, to some extent, has affected the development of teacher education in Senegal.

The educational degrees were recognized by an accredited authority of the French Ministry of Education. Moumouni (1964) noticed that,

> Just as we need interpreters to help us understand the natives, as well, we be intermediaries belonging to indigenous communities in their origin and the European environment by their education, to make people understand the country and make them adopt this foreign civilization for which they show, without one can hold it against them, a misoneism very difficult to overcome.

The early types of education systems implemented at all levels were developed through the following government actions. The first action was to develop and implement a comprehensive program of school construction that led to the creation of general secondary schools (CEG) in all regional and district capitals.

For the second action, the State did recruit young teachers, low level of general education (5th, 4th and 3rd high school) and insufficient professional qualification (accelerated teacher training during an internship of two to three months during long school holidays) and that formed in the history of the school, the generation of educational instructors.

These two combined actions have led to a share-related effect, the increase in capacity at the middle school, and therefore the increase in staff, and also a significant increase in the rate of enrollment from 1960 to 1970.

Organization Administrative and Educational Sequences

In Senegal, ministries and agencies are often reorganized. A notice issued in 2010 established three ministries responsible for governing education: the Ministry of Pre-School, Elementary and Secondary Education, and National Languages; Ministry of Technical Education and Vocational Training; and Ministry of Higher Education, Regional University Centers and Scientific Research (UNESCO 2010). Also, as in many other developing countries, a 10-year education training plan (PDEF: *Programme Decennal de l'Education et de la Formation*), which aims to improve the state of education, has been drawn up.

The sharp increase of elementary education enrollment in 2011 is credited to these reforms. In 2012, a new education plan *(Programme de Dévéloppment de l'Education et de la Formation* 2012–2025) was drawn up, which included the reorganization of the departments (Taniguchi, 2014).

Non-formal education

The non-formal education sector, under the responsibility of the Ministry of Technical Education, Vocational Training, Literacy and National Languages, includes literacy, basic community schools, and Franco-Arab schools. The development of this sector must necessarily be based on the quality of available human resources.

Formal education

This sector covers many levels and types of education. It is composed of preschool education, elementary education, secondary education and general secondary education (the 6-4-3 years education system), technical education and vocational training and higher education.

Elementary education

Elementary education is intended to instill "the basics" in children 7 to 12 years old: reading, writing, numeracy, knowledge of the environment, useful knowledge and skills required for living in community and prepare access to the upper levels. Over a period of six years, it is divided into: Basic Course (CI), Preparatory Course (CP), First Year Elementary Course (CE1), Elementary Second-Year Courses (CE2), Medium First-Year Courses (CM1) and Second Year at Intermediate (CM2).

The teaching of the Arabic language is taught as optional for four years. For the recruitment of students, the government gives priority to children seven years old. The youngest (age six) are admitted within the limits of available places, if they have completed pre-school (Ministry of Education, 2003).

A Final Certificate of Elementary Studies (FEAC) is recommended at the end of the six years of elementary school. The same test is also used for access to middle school.

Middle school education

It is provided in the means of colleges (EMC) whose cycle lasts four years, from 6th to 3rd. It is an extension of elementary education. This level lead to the Diploma of End of Middle School Studies (BFEM).

Secondary general education

General secondary education includes three years of study: the Second, Premiere, and Terminal. In 1999–2000, 53.8% of students who completed middle school entered this phase. The secondary general education offers two sets of studies: a literary series "L", with two options (L1 or L2) as foreign languages and scientific series "S" with two options (S1 or S2) to bring together the economic and experimental sciences or mathematics. These series replaced the oldest series A, D, C, E, F and G in 1987. The studies lead to the Baccalaureate degree.

Technical secondary education

An education in technical colleges and vocational schools. It is provided by the Department of Technical Education, Vocational Training, Literacy and National Languages.

Higher education

Universities and training institutions covering areas of increasingly diversified knowledge. These universities and institutes are the highest degrees and maintain relationships with foreign universities of Western Europe and North America. They welcome students from around the world. The higher education accredits a License degree, a Master degree and a Doctoral degree.

Weakness of the primary and secondary school mathematics

Teaching weaknesses begin in the primary schools. Recruitment of teachers is a determinant factor to ensure that students have access to high quality mathematics education. However, recruiting mathematics teachers is not easy because few students want to be mathematics teachers. This explains the shortage of teachers.

At every level of education, more students drop out of mathematics. In addition to low primary school enrollment, there are high dropout rates between primary and secondary school, during secondary school, and between lower secondary and higher secondary school.

Secondary school education suffers from a lack of teachers with mathematics training, as well as low student enrollment. Research has also found that teacher quality in Senegal is inadequate (Michaelowa, 2001). Teachers may be hired quickly as volunteers for short-term positions to fill gaps in the faculty, and end up with a permanent position without any training in classroom management and instruction. This contributes to poor performance in mathematics. Teacher development and teaching seminars are rare due to limited funding, and are not structured enough to impact teaching practice when they do occur. A report from the International Mathematics Union (IMU) shows that "After 1980, low public budgets in sub-Saharan francophone countries forced the reduction or even elimination of funding for teacher-training programs and recruiting of fully qualified teachers."

Identifying exceptionally gifted mathematics learners is crucial in the sense that it ensures that mathematically gifted students reach their maximum potential to the benefit of their countries or themselves. Unfortunately, Senegal has few structured systems for identifying highly gifted mathematics students. The most common method of identifying mathematically gifted learners is through their participation in the Pan-African Mathematics Olympiads (PAMO) founded by the African Mathematics Union (AMU) to encourage young talents in mathematics and to exchange information on curricula and methods.

In Senegal, screening for talented pupils in mathematics starts at age 15. National competitions are organized and the best students are followed, trained and selected for the PAMO or International Mathematics Olympiads (IMO). Additionally, a national competitive examination at the bachelor's level is organized and winners are awarded excellence scholarship to study abroad. Usually opportunities depend on the socioeconomic background of the students. A gifted child in mathematics, whose parents cannot afford fees, may not attend school at all. A gifted student whose parents are wealthy may seek advanced educational opportunities either in France or American universities.

Teaching jobs are often low in prestige and pay. Therefore, most highly gifted students in mathematics leave the field for more rewarding opportunities such as telecommunications, engineering, information and communication technology.

Weakness in mathematics at the higher education level

Senegal is among the list of African countries that have a strong university education. University mathematics includes algebra, differential geometry, partial differential equations, applied mathematics (modeling), probability/statistics. For a population estimated at 12.5 million, Senegal has a total of seven universities. The number of doctorate holders in mathematics is estimated to 67 and the number of math doctoral degrees awarded is 17.

As aforementioned, Senegal's education weakness begins in the primary schools with a lack of qualified teachers and low student participation. Those few Senegalese students who enroll in universities to become mathematicians encounter many difficulties such as poor infrastructures, outdating teaching methods and so on.

For example, classrooms originally built to hold 30 to 40 students host nowadays several hundred students. Many stand along walls or skip class entirely when they cannot gain entry. Even those who arrive early enough to find a seat have difficulties taking notes, asking questions, and finding tutorials or advisors. In addition, many buildings were built during the colonial area of 1960s and have inadequate electrical wiring. Students seldom have access to computers, internet access, textbooks or journals. The few operational public computers are kept in small labs and few students can afford their own. Libraries built decades ago are expected to serve several thousand students today.

Similarly, teaching does not map with students' career needs. Curricula often include no career guidance for students who simply finish as a "math major." Universities tend to emphasize either pure mathematics or applied mathematics, but seldom both. Emphasizing pure mathematics does have advantages such as preparing students for deeper understanding and for a variety of application. More importantly, pure research develops a rigor of thinking and assessment. However, an overemphasis on pure mathematics may map poorly with students' career needs. Career vision is usually limited to university teaching, which attracts only a few, leading to a high dropout rate.

Programs with strong emphasis on applied mathematics well prepare students to apply their skills in government, industry or multidisciplinary work but lack strength in pure mathematics.

On the other hand, low numbers of full professors and heavy teaching loads affect faculties as well as students. While students receive too little attention, most faculties have too little time to advance in their career through research and other activities. Because of limitation of resources many faculties are unable to attend professional conferences and are poorly paid compared with their counterparts in government and the private sector.

One consequence of the poor teaching conditions is that institutions are unable to fill faculty jobs. Many lecturers have received Masters and PhD degrees from European and American universities. However, when they return home, the lack of career support affects their enthusiasm and energy. Their interest in research evaporates as they work part time jobs to supplement their low compensation.

Another weakness results from low image of mathematics and insufficient public supports. For example, common misconceptions are that mathematics is for only a few selected students or only for male students. Many students have difficulties seeing the value of mathematics in real life and thus have a low opinion of mathematics as a career.

Overall, Senegal, like other African countries, have talented students with the potential to succeed in mathematics. However, in the face of difficulties described above, mathematics majors from good universities must do so with very little support.

Features of strengths

Recently, the previous government raised its commitment to high education in general, including mathematics. The compensation for university professors has increased, making academic jobs more attractive to many young Senegalese who wish to return home after earning a PhD abroad. Furthermore, Senegal regularly participates in the Pan African Mathematics sponsored by the African Mathematics Union (AMU) to encourage youthful talent in mathematics and to exchange information on curricula and teaching methods. In addition, the West African Training School (WATS) in Saint Louis Senegal is considered to be especially strong. It has been designated an "affiliated center" by the Office of External Affairs (OEA) of the ACTP.

Another support program is the creation of researches and education centers of excellence to strengthen mathematics education in the country, for example: FASTEF, IREMPT, IGEN, INEADE. Cheikh Anta Diop University in Dakar has a strong center specializing in Algebra lead by

Professor Sangare with a small group doing differential geometry. Gaston Berger University in Saint Louis has a group specializing in partial differential equations and applied mathematics (modeling).

Furthermore, Senegal benefits from the small multi-country networks in Francophone Africa that offers many exchange opportunities as well as coursework and conferences. Areas emphasized at some institutions might be considered old-fashioned (algebra analysis). However, other institutions are expanding into newly relevant area.

Other support programs are the African Mathematical Union (AMU), whose current activities are seen to be confined to support for the Pan-African Mathematics Olympiad in which several African countries participate. Similarly, the Agency of Francophone Universities (AFU) is a worldwide network that supports conference travel, doctoral scholarships, cooperative projects, a center of excellence, networks and other activities in Africa.

There is also SARIMA (*Soutien aux activities de recherches informatiques et mathematique en Afrique*), which supports regional centers in Senegal. SARIMA was funded from 2004 to 2008 by the French minister of foreign affairs. Their emphasis is on information sciences and applied mathematics.

These programs make a generally positive impact that includes improvement in universities facilities and infrastructures, enhanced training of mathematics teachers and instructors, and bursaries for talented students who could not otherwise afford postgraduate education. There are encouraging signs, such as talented young African mathematicians who have found productive careers in many countries and are willing to help in developing capacity in their home country. Also, increasing numbers of mathematics books are being published by nationals of Senegal. The internet is accessible to mathematics students, allowing them access to journals, problems and online courses. The most encouraging sign is the creation of an institution of doctoral studies in mathematics and informatics in Senegal that forms and awards Masters and PhD degrees at home. There is no more need to send promising students abroad for post graduate education.

Conclusion

The story of mathematical development in Senegal is one of potential unfulfilled. Based on the outstanding achievement of some individuals and institutions, it is clear that Senegal has no lack of talented potential mathematicians. However, without a strong educational structure at all levels reaching that potential is not easy.

Strengthening mathematics development in the country requires simultaneous supports at all levels, from elementary school to postgraduate research. To concentrate on primary education alone will be futile if there are no qualified teachers or skilled mentors. More supports are needed from the government for those who are willing to become educators or researchers in mathematics. For example, improving infrastructures to include expanded internet and computing facilities would allow better linkage with resources and the academic community, providing scholarships to graduate students and fellowships for faculties and a clear path to rewarding mathematics based careers.

In addition, more research centers and more scholarship opportunities for postgraduate studies are needed. Also, mathematics clubs need to be established and promoted in secondary school to help develop students' participation in learning and problem solving.

References

Developing Countries Strategies Group International Mathematical Union: *Mathematics in Africa, Challenges and opportunities, 2009.*

Joseph Gaucher: *The beginnings of education in Francophone Africa, Jean Dard and mutual school of Saint-Louis-du-Senegal, the African Book, 1968.*

Mamadou Sanghare: *Challenges of Teaching Mathematics in Senegal.*

National Directory school year statistics (2007–2008), University Office of Statistics and School Documentation.

Michaelowa, Katharina, 2001. *"Primary Education Quality in Francophone Sub-Saharan Africa: Determinants of Learning Achievement and Efficiency Considerations," World Development, Elsevier, vol. 29(10)*, pages 1699–1716, October.

Ministry of Education national, Senegal (2003): *The Education and Training program (Education for all).*

Moumouni, A. (1964): *Bulletin de l'education en AOF, no. 3, juin 1917.*

Ritsu Taniguchi, *Postdoctoral Research Fellow of the Japan Society for the Promotion of Science (JSPS).* January 24, 2014.
UNESCO (2010): *Science policy, and capacity-building. Science Report, 10, November, Paris.*

About the Authors

Mr. Serine Ndiaye is a PhD student at Teachers College, Columbia University and is Mathematics Instructor at Borough of Manhattan Community College. He was born and raised in Mbacke, a city in Senegal, where he received a Bachelor's of Sciences from Cheikh Anta Diop University. In 2001, he obtained a Master's degree in Urban Economic from University of Missouri in Kansas City and completed a Master's degree in Mathematics Education from Teachers College, Columbia University in 2014. He is the principal author for the Senegal chapter "Weaknesses and Strengths of Mathematics Education in Senegal." His research interests are Developmental mathematics, Problem Solving and Mathematics Literacy.

Bakary Sagna is a doctoral student in the mathematics education program at Teachers College Columbia University. He received his MS in Education and BA in Mathematics from Lehman College in New York City, and his BST in Mechanical Manufacturing from Senegal's University Check Anta Diop in Dakar. Bakary currently teaches mathematics at CUNY Hostos College and Monroe College in New York. Fields of interest include Health Services and Radiology, Statistics and Algebra. Bakary is also co-editor of algebra handbooks for CUNY Proficiency Exams at CUNY Hostos. His research interests include strategies in teaching and learning mathematics at the secondary level, and the use of technology in mathematics and science instruction.

https://doi.org/10.1142/9789813146785_0012

Chapter XII

SUDAN: The Development of Mathematics and Mathematics Education in Sudan

Mohamed El Amin A. El Tom
The Ministry of Education, Khartoum

Introduction

A proper understanding of an education system presupposes a knowledge of the broad context in which the system functions. The main purpose of this introductory section is to outline such a context. Modern mathematics is rather young in most developing countries, including Sudan. In the next section, I present major efforts at building capacity in mathematical research and mathematics education. The major developments in the education system since independence are considered in section 3. A number of opportunities for mathematics curriculum reform have been presented over time, some of which were missed. These are considered in section 4. Aspects of mathematics teacher preparation and training are discussed in section 5. A brief consideration of internationally visible research carried out by Sudanese in both mathematics and mathematics education is presented in section 6. Some concluding remarks are made in the final section.

Diversity is one of Sudan's defining characteristics. With an area of about 1.9 million square km, Sudan is the third largest country in Africa (after Algeria and DR Congo) and the 16th largest in the

Figure 1.

world. It was the largest country in Africa until 2011 when South Sudan seceded. Its population in 2015 is estimated at 40 million, giving rise to a population density of about 22 per square km. It has seven neighbors (Central African Republic, Chad, Egypt, Eretria, Ethiopia, Libya, and South Sudan). Since the eleventh century, the route through Darfur (in the West) to Mecca was the favorite one for many West African pilgrims. Many of these West Africans either never made it to Mecca or they settled down, particularly in Darfur on their way back. It is said that the largest group of West African immigrants came into Sudan during the colonial period (al-Nagar, 1972).

Except for the Nile(s), a flood region (10%), and a mountainous terrain (1%), the country comprises two halves: one desert (29%) or semi-desert (19%) and the other savannah with varying degrees of rainfall. (See UNEP 2007).

Demographic Characteristics

Sudan is a multicultural, multi-linguist and multiethnic country undergoing rapid change. Located at the crossroads of North, West and East Africa and the Middle East, it has a number of ethnic and language groups, including the Shaigia, Fur, Zaghawa, Nubians, Beja, Rashaida, Manassir, and Fallata.

The majority of the population is rural (64.4%). However, there are important variations between states (18) in this regard. For example, whereas the proportion of inhabitants living in rural areas in Khartoum state is 19.9% only, it is 45% in Red Sea state and 82.3% in Northern state. But, it must be noted, that the country is undergoing rapid urbanization. Thus, the proportion of the urban population increased from 27% in 1990, to 34% in 2014 (retrieved on 12/02/2015 from http://www.tradingeconomics.com/sudan/rural-population-wb-data.html).

Sudan has a young population, with some 40% under age 15; about 20% are between ages 15 and 24 years. Moreover, this young group grows at a fast rate estimated at 2.4% per year (2010–2015). This means that there is a big and growing demand on education. Also, it is expected that the age cohort 5 to 14 years old will grow at an annual rate of more than 16% during the 5-year period 2013–2018. A simple computation shows that the size of this age group stands at about 9.96 million in 2015 and, according to Population Pyramid (http://populationpyramid.net/sudan), it is projected to be 10.94 million in 2020. Clearly, this demographic pressure has important implications for educational policies: in addition to having to achieve the goal of education for all (EFA) (the gross primary enrolment rate is 69.8% in 2011–2012), it is necessary to meet the educational needs of about one million additional kids in 2020.

Political Developments

Two salient features have dominated Sudan's political landscape for the past 50-odd years: civil strife and authoritarian rule. During its 53 years of independence, the country was under authoritarian rule for 42 years (1958–1964; 1969–1985; and 1989 to date), which means that all living

Sudanese have either no solid experience or no experience at all of democracy.

After the first phase of civil war in the south (1955–1972), war broke out again in 1983. However, this time around it progressively engulfed larger areas of the country: Nuba mountains in southern Kordofan, southern Blue Nile, and East Sudan. This second phase continued unabated for 20 years. A historic Comprehensive Peace Agreement (CPA) was signed between the government (National Congress Party) and the Sudan People's Liberation Army/Movement (SPLA/SPLM) in January 2005.

Sadly, just before the guns of Africa's longest-running war were completely silent, war broke out in Darfur, western Sudan, in 2003. And despite numerous national, regional, and international peace initiatives, peace in Darfur remains elusive.

Economic Development

According to the World Bank, Sudan's economy was classified as lower-middle-income in 2014 (GNI per capita of $1,045 to $4,125). Indeed, Sudan's GDP (PPP) per capita during the period 2009–2013 has fluctuated between a low of $2,151.61 in 2009 and a high of $2,682.99 in 2011 (see Table 1). The country has failed to achieve the required rate of growth for sustainable development (more than 6%) during the past 25 years (1980-2014). As Figure 2 shows, the average annual rate of growth of the economy has remained below 5% for most of those years. The oil sector was the main factor behind the significant improvements in economic growth since the country started exporting oil in 1999. Moreover, whatever rate of growth has been achieved, it has not translated into wealth for the average Sudanese (see Figure 3).

Agriculture is the mainstay of Sudan's economy, accounting for 27.4% of GDP, and continues to employ about 80% of the work force. Although the country has about 200 million acres of arable land, it exploits 20% only. It is noteworthy that traditional agriculture exploits 60% of cultivated land and employs 65% of farmers.

Table 1. Sudan: GDP (PPP) per capita, 2009–2013.

2009	2010	2011	2012	2013
$2,151.61	$2,178.85	$2,682.99	$2,544.63	$2,550.10

Source: IMF World Economic Outlook (WEO), April 2015

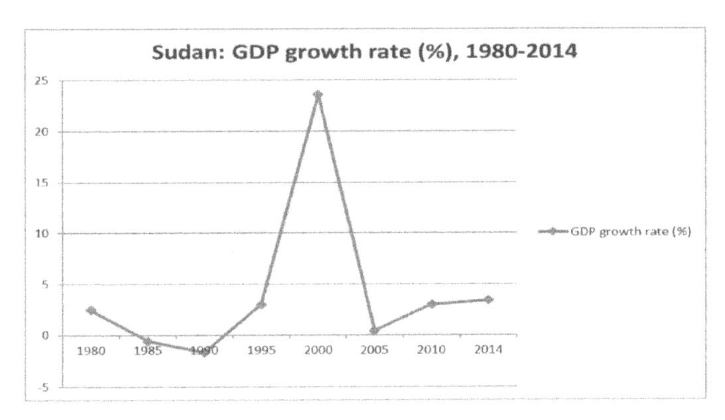

Figure 2. Sudan: Real GDP growth, 1980–2014.
Source: IMF World Economic Outlook (WEO), April 2015

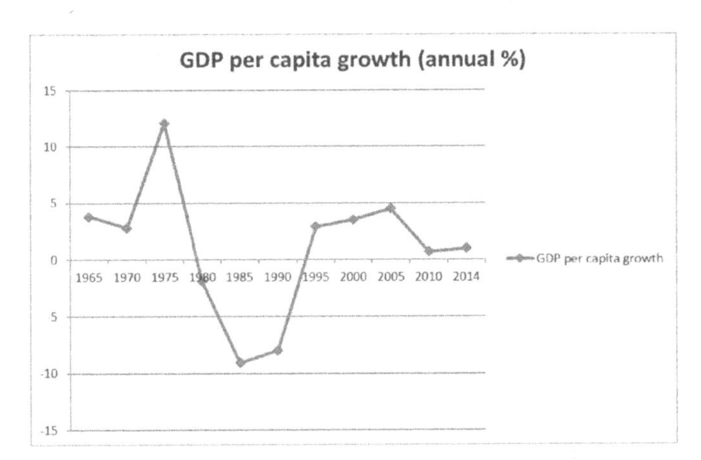

Figure 3. Sudan: GDP per capita growth (annual %), 1965–2014.
Source: IMF World Economic Outlook (WEO), April 2015

The economic development of Sudan is partly hampered by the low level and poor quality of its stock of human capital—that is the skills available in the population and the labor force. Educational attainment is a commonly used proxy for the stock of human capital. The low educational attainment of the adult population is clearly demonstrated by the data in Table 2. To see these data in proper perspective, it is instructive to compare them with data for the Republic of South Korea, a country whose GDP per capita in 1960 was comparable to that of Sudan and whose success in achieving sustained significant levels of development is largely due to its investment in human capital (see Table 3).

Table 2: Sudan: Educational attainment for population aged 25 and over, 2000 (5) 2010.

	Units	**2000**	**2005**	**2010**
Percentage of no schooling attained in Pop	%	62.5	58.8	55.7
Percentage of complete primary schooling attained in Pop	%	19.8	24.1	26.0
Percentage of complete secondary schooling attained in Pop	%	4.6	3.9	3.9
Percentage of complete tertiary schooling attained in Pop	%	0.8	1.2	1.8
Average years of schooling attained	Years	2.44	2.93	3.13

Source: Barro, Robert and Jong-Wha Lee, April 2010, "A New Data Set of Educational Attainment in the World, 1950-2010." NBER Working Paper No. 15902.

Table 3. South Korea: Educational attainment for population aged 25 and over, 1960 (5) 1970.

	Units	**1960**	**1965**	**1970**
Percentage of no schooling attained in Pop	%	56.9	43.6	34.3
Percentage of complete primary schooling attained in Pop	%	26.2	33.5	36.5
Percentage of complete secondary schooling attained in Pop	%	5.8	7.6	9.9
Percentage of complete tertiary schooling attained in Pop	%	1.9	2.7	4.3
Average years of schooling attained	Years	3.12	4.26	5.20

Source: Barro, Robert and Jong-Wha Lee, April 2010, "A New Data Set of Educational Attainment in the World, 1950-2010." NBER Working Paper No. 15902.

Social and Human Development Context

Socio-economic development has been greatly affected by civil strife and related governance problems. "With increased spending on defense and security as a result of armed conflicts, budgetary allocation to infrastructure, health and social services has dwindled. Conflict and internal displacement of civilians have resulted in food insecurity in parts of the country and continue to cause egregious suffering and loss of life" (African Development Bank, 2012). 46% of the population is considered poor. Sudan ranks 166 out of 187 countries as per the 2014 Human Development Report and has "alarming" levels of hunger according to the 2013 Global Hunger Index (WFP 2015).

Sudan is classified as having insufficient progress in achieving the fourth Millennium Development Goal (MDG-4), reduce child mortality, as levels of child and infant mortality are among the highest in the region and the world (SHHS 2010). The data in Table 4 show that the levels of underweight, stunted and wasting children under 5 years are high.

Table 4. Sudan: Proportion of underweight, stunted and wasting children under 5 years(%).

Condition	Global and state with highest and lowest proportion (%)			
	Global	**Highest**	**Lowest**	**(UNICEF) Definition**
Underweight	32.2	Sinnar: 42.6	Khartoum: 19.9	Moderate and severe—below minus two standard deviations from median weight for age of reference population. Severe—below minus three standard deviations from median weight for age of reference population
Stunted	35	Red Sea: 54	Khartoum: 21.9	Moderate and severe—below minus two standard deviations from median height for age of reference population
Wasting	16.4	Red Sea: 28.5	Khartoum: 12.8	Moderate and severe—below minus two standard deviations from median weight for height of reference population.

Source: SHHS (2010)

These contexts demonstrate that Sudan faces three major mutually reinforcing challenges:

- The achievement of a durable peace
- The achievement of a democratic and just society
- The achievement of sustainable development.

The role of education, mathematics, science and technology in addressing these challenges effectively cannot be overemphasized.

Development of Mathematics and Mathematics Education: Major Efforts

At independence, in 1956, Sudanization of the civil service was a top priority for the government. Likewise, the administration of the University of Khartoum, the only national university in the country, launched a grand scheme of scholarships for all its distinguished graduates to pursue graduate studies in their respective disciplines abroad, mostly in the UK. However, mathematics presented a special problem in this regard. For, not only was there no Sudanese with a postgraduate qualification in the discipline, but able students in mathematics chose overwhelmingly to study engineering. In 1961, the university offered two scholarships for students who distinguished themselves in the second year final examinations in mathematics to study for a bachelor degree and, if they achieved the required graduation standards, to continue for a doctorate degree in British universities. A total of five students earned their PhDs through this scheme and returned to work in the university. The scheme was discontinued in 1965.

From this modest beginning some visible progress was made over the years, but there were also setbacks. Realizing that mathematics was organized in three different Faculties, the first of the five returnees proposed to the Vice-chancellor in 1969 that resources should be pooled into an autonomous School of Mathematics. Also, he organized a regular research seminar in mathematics and convened a small group to discuss issues of mathematics education in Sudan. The deliberations of this group led to the organization of the first International Conference on "Developing Mathematics in Third World Countries" (El Tom, 1979).

In 1978, the newly established School of Mathematical Sciences could boast more than 25 faculty members, including 14 Sudanese PhDs in mathematics, and offered six bachelor's degree programs (mathematics, statistics, computer science and a double degree in any two of the three). In 1980, the School organized a one-year symposium in "Mathematics and Development" (see, for example, Nasell 1985 and Schweigman 1986 for two of the proceedings).

Within a decade of its existence, the School managed to establish itself as an institution of excellence within the university and as the leading institution of mathematical sciences in the country. Unfortunately, the worsening economic situation throughout the 1980s led to the exodus of most of the School's faculty and by 1994 there were only four faculty members left! The Department of Mathematics, University of Liverpool, UK, cancelled the Agreement of Collaboration with the School out of concern for human rights in Sudan following a coup d'état in 1989. The fervor of both faculty and students of earlier years seemed to have evaporated and the whole project of building mathematics in the country seemed to have collapsed under the weight of economic and political developments.

The School was promoted to Faculty status in 1996. Today, the faculty includes 27 Sudanese PhDs (12 each in mathematics and computer science and three in statistics) among its 65-strong faculty, offers six bachelor degree programs (mathematics, statistics, computer science, information technology, statistics and computer science, and mathematics and computer science). Information Technology is by far the most competitive and popular of the six programs. Also, the school offers five masters degree programs (pure mathematics, industrial mathematics, statistics, computer science and information technology).

What about Sudanese specialists in mathematics education? The University of Khartoum has never made an investment in building capacity in mathematics education. The few Sudanese who earned a doctorate in the discipline, three in all, have financed themselves. Only one of them is working in the country at present.

Of course, the higher education institutional landscape of mathematics is not limited to the University of Khartoum. Indeed, the system of higher education in Sudan witnessed a huge expansion during the 1990s. In 1989, the system was comprised of a total of 17 institutions, including four public universities, two private universities and an Egyptian university. In 2013, the system had 108 institutions: 31 public universities, 56 private institutions (13 of which are universities), 20 public technical colleges and an Institute for Translation. Of the 108 institutions, only some public universities offer bachelor degree programs in the mathematical sciences and/or bachelor of education degrees in mathematics. In fact, of the 31 public universities, only six offer bachelor degree programs in the mathematical sciences (four in mathematics); and while all of them have faculties of education, only nine offer a bachelor of education (B. Ed.) degree in mathematics, ten in mathematics with physics and two in computer science.

Besides the University of Khartoum, two of the four universities which offer bachelor degree programs in mathematics, also offer master degree programs in mathematics. The total number of faculty members with an earned PhD in mathematics in the four universities is about 45.

None of the 31 faculties of education have a faculty member holding a doctorate degree in mathematics education (faculty lists of public universities, 2014–15).

The Educational System in Sudan: Major Developments

Undoubtedly, the establishment of a modern system of education in Sudan in 1900 represents the most important social project implemented by the condominium during the entire period of colonial rule. The system exhibited four important features:

1. A 4-4-4 (primary-intermediate-secondary) educational ladder;
2. A nationally mandated curriculum;
3. The medium of instruction was Arabic for the first two stages of the general education cycle and English for secondary and post-secondary education; and
4. A national institute of education for, among other things, curriculum development and training of primary as well as

intermediate school teachers (established in 1934 at Bakht er Ruda in central Sudan).

Not unexpectedly, all of these features have changed over the years in significant ways except feature 2, the nationally mandated curriculum.

Shortly after the overthrow of Aboud military regime in a popular upheaval in 1964, an important policy measure was implemented, namely the medium of instruction in secondary schools was changed from English to Arabic in 1965. Various subject matter groups were charged with the task of translating the existing curricula from English into Arabic.

The second important change in the colonial system of education occurred shortly after the pro-Arab nationalism coup d'état of 1969. In 1970, the government decided to implement the educational ladder adopted by most Arab countries, namely a 6-3-3 (primary-lower secondary-upper secondary) ladder. In 1990, shortly after the 1989 coup d'état, the school cycle was shortened by one year and changed into an 8-3 (basic-secondary) cycle. Also, since 1992, pre-school education has become, for the first time in the history of Sudan, part of general education.

Bakht er Ruda Institute of Education was founded in 1934 as a training college for primary-school teachers. By 1984 and after 50 years, the Institute developed into an entity much greater than a usual setting of an institute of education. It occupied an area of 400 acres, including farms of about 200 acres. It housed the Curriculum and Books Center; the School Supervision and Field Teachers' Guidance Office; the National Center for Educational Research, and an audio-visual aids unit; the Institute of Education for primary-school teachers (IEPST); the Institute of Education for Intermediate school teachers (IEIST), which had 13 departments; the center for the training of teachers of the Integrated Rural Education Centers (IREC's); three intermediate schools; one primary school; one integrated rural education center; a central library; two nursery schools; in addition to the general secretariat (administration, finance, personnel, stores, public relations and follow-up). In 1996 the Institute was dismantled and what remains of it today is the National Center for Curriculum Development and Educational Research, part of the Federal Ministry of Education.

Sudan is a federal state and the responsibility for general education is shared, since 1993, among the basic administrative units of the federation (Center, states and localities—subunits within each state). The responsibilities of the Federal Ministry of Education include formulating general policies, plans, programs, and curricula for academic and technical education, and training of teachers and educational administrators. States, on the other hand, are responsible for the management and funding of general education.

I end this section by considering three salient features of the education system.

Rote Learning

Anyone who attended a Sudanese school in 1960 and observed a class today would notice no difference in the behavior of teachers and learners in the two classes that are separated by more than half a century. Although research has provided significant insights into how we learn and how we teach over the past decades, Sudanese schools have stuck stubbornly to their outdated practices in this regard. This is because of deeply rooted beliefs that learning can only occur in a particular way. This applies even to university Faculties of Education which are responsible for the preparation of teachers for general education. Thus, rather than emphasizing ways of thinking which involve creative and critical approaches to problem-solving and decision-making, education in Sudan has become about reproducing content knowledge. Of course, routine cognitive skills are the easiest to teach and easiest to test.

Excessive Obsession with Examinations

Modern education was introduced during the condominium period primarily for the creation of a male elite trained for administrative service. Sudanese occupying government jobs were accorded a privileged social status. Elsewhere (El Tom, 2006, p. 29), I observed that "Students, parents and society at large view schooling as their principal means out of poverty and/or their gateway to wealth and power through government employment." I further noted that the "singular importance accorded to government employment is as old as formal education itself. Thus, a commission of inspection of Gordon Memorial College (precursor of University of Khartoum) . . . noted in its report that ‘‘the general aim of the Gordon College higher schools is to give an education of a secondary standard to boys of sixteen to twenty years of age which will fit them for posts of various kinds in government offices and commercial houses. The great majority of pupils have so far preferred the former” (cited in El Tom, 2006, p. 29).

With time, the main, and to some the only, purpose of education in Sudan has become attaining a standard in examinations in class or at the end of a stage that would permit the student to pass on to the next class or to the next stage. Examinations have become the driver of the teaching and learning process. The examination itself, whether at grade level or at the end of a stage, has become stereotypical. Books that contain end of stage examinations together with model answers are published annually and may be bought from most bookshops and in many cases from kiosks. Any part of the syllabus that is not covered by these stereotypical examinations would not be taught.

Thus, the phenomenon of private lessons has become widespread, especially in cities. The explicit purpose of such lessons is to prepare the student for the examinations. And since these lessons are very expensive, many schools offer them for groups in order to reach students of low socioeconomic status.

During the last six months of an academic year, most schools require their final grades students (G8 for the basic stage and S3 for secondary stage) to sit for monthly tests, join camps where they are drilled for the examinations and take mock examinations. Students in private secondary schools who don’t perform well enough in mock examinations are not allowed to sit for the Sudan Secondary School Certificate examinations where they are attending school because these schools wish to achieve 100% success rate.

At the end of every academic year various forms of celebrating success at the end of basic and secondary stage are exhibited. Some schools publish in widely-read national newspapers news of their excellence and photographs of their distinguished students. The state announces, at the level of the Federal Ministry of Education and the Ministry of Higher Education and Scientific Research, the results of the Sudan Secondary School Examinations in a news conference that is televised and broadcast in prime time nationally. Also, families celebrate the success of their daughters and sons not only at the level of the Sudan Secondary School Certificate, but also at the level of kindergarten.

It is with some sadness that one notes that in these celebrations no one asks: What have these students learnt?

Assessment

As indicated in the previous section, the majority of stakeholders in education in Sudan have long believed that the main purpose of education is certification. Under such circumstances, the most relevant method of measuring students' achievement is summative assessment. Other types of assessment that are known to be effective in improving students' achievement, such as formative assessment, are not considered by the system for adoption either because they are: i) unknown to teachers; or ii) may disrupt the system (formative assessment, for example, could be demanding on teachers).

In the case of mathematics, the continued widespread use of summative assessment is contrary to evidence from research and practice of greater student achievement in classrooms where teachers use formative assessment (Black & Wiliam, 1998). Also, in their meta-analysis, Ehrenberg et al. (2001) report the impact of formative assessment on student achievement being four to five times greater than the effect of reducing class size.

School Mathematics: Major Opportunities for Curriculum Reforms

School mathematics curricula remained unchanged for nearly twenty-five years following independence. But, as I will document later in this section, a number of opportunities for mathematics curriculum reform presented themselves and some were missed.

A characteristic of the primary and lower grades of the intermediate schools' mathematics curricula was their focus on arithmetic operations.

The two main textbooks for these grades (G1–G6) were titled *Pupil's Arithmetic* and *Dictation Arithmetic*.

The first opportunity to undertake a significant reform of the general education curricula arose in 1970 when the educational ladder was changed. Prior to that, the change in the medium of instruction in secondary stage provided an opportunity for reform, which was more or less missed.

Arabization of the Medium of Instruction in Secondary Schools

For nearly a decade after independence, the secondary school mathematics curriculum was based on the well-known trio of Durell (see the Appendix for a sample of contents):

1. *School Certificate Algebra* (first edition 1937)
2. *A New Geometry for Schools* (first edition 1939)
3. *General Arithmetic for Schools* (first edition 1936).

One would have expected the decision in 1965 to change the medium of instruction from English to Arabic to give rise to a major curriculum reform. However, as mentioned in the previous section, all that was done in the case of mathematics was to ask a few experienced mathematics teachers to write textbooks based on translation of selected topics from Durell's trio into Arabic. Thus, the old curriculum remained largely in force, albeit in Arabic.

Change in Educational Ladder

With the change of the educational ladder from 4-4-4 to 6-3-3 (the ladder adopted by most Arab countries, notably Egypt) in 1970, an opportunity for major curriculum reform presented itself. However, the Ministry of Education adopted the simplest of all viable courses of action, namely shifting the curriculum downwards. Thus, the curricula for grades 5 and 6 of the new primary school are the corresponding curricula for grades 1 and 2 of the old intermediate school, respectively; and the curriculum of grade 9 of lower secondary is the corresponding curriculum of grade 9 (S1) of the old secondary school.

Influence of the 'Modern Mathematics' Movement

In 1969, UNESCO initiated a project for the development of mathematics curricula in Arab countries. Regional and national workshops were organized for this purpose. In Sudan, Bakht er Ruda Institute embarked on the preparation of curricula for the lower secondary stage (G7–G9), which were clearly influenced by UNESCO project. The fruit of this effort saw the light in 1976 in the form of a textbook for G7 including, for the first time, modern mathematics concepts such as sets, groups and rings. However, efforts at modernization of higher secondary school (G10–G12) mathematics curricula remained unimplemented until they were overtaken by political events in 1989 when a military coup displaced the existing regime at the time.

1990 Educational Ladder

Following the change of regime in June 1989, a new educational ladder: 8-3 replaced the 6-3-3 one in 1990. The new ladder has reduced the duration of the lower school stage by one year and integrated the resulting stage into basic education. In addition, the planned school year of 250 days is, in practice, 180 days long only. This means that the present basic education stage lasts for a total of 1440 days, which is less than the duration of the previous duration of the combined primary and lower secondary stages by a full 810 days! In other words, in practice nine years of education have been reduced to 5.76 years, each lasting 250 days. The impact of these features on the curriculum has been significant.

In the first two years of secondary education, students follow the same curriculum; in the final year (G12), students choose between the arts and science streams. Prior to the year 1990, two mathematics courses were offered in G12: "general mathematics" for all G12 students and "additional mathematics" for a relatively small number of students who are judged to be able in mathematics. The latter course, which was compulsory for students wishing to enroll in bachelor degree programs in mathematical sciences or engineering, consisted of two parts with separate textbooks: a) pure mathematics (some algebra, analytical geometry, differentiation and integration; and b) mechanics. As time passed, many of the geometry topics in the "general mathematics" curriculum have been removed and the mechanics part of the additional mathematics course disappeared (with many of its components added to a physics course).

Today, and since 2003, three different courses are offered in G12: basic mathematics for the arts stream; and Specialized Mathematics: Book 1 and 2 for the science stream. Basic mathematics is comprised of five units: real functions and limits; differentiation and integration; statistics; probability; and matrices. The units of specialized mathematics are presented in Table 5.

Teacher Preparation and Training

In 1993, the pre-service qualification of basic education teachers was revised from a two-year teaching diploma to a four-year bachelor of education (B. Ed.) degree. The two-year diploma course was offered by a network of 73 in-service education training institutes (ISETI) across (northern) Sudan. Teachers attended the institutes one day per week and taught in the schools for the remainder of the work week. Funding for the ISETI diploma ceased in 1993, and the staff of the larger institutes was absorbed into the universities' faculties of education, which would be responsible for qualifying basic education teachers through a B. Ed. Degree, which is a composite course of general and professional education.

Table 5. The main units of the mathematics curriculum for G12, 2014–2015

Unit	Specialized Mathematics: Book 1 (Arts track)	Specialized Mathematics: Book 2 (Science track)
1.	Mathematical deduction – Permutations & combinations – Binomial theorem	Real functions & limits
2.	Matrices	Differentiation
3.	Partial fractions	Applications on differentiation
4.	Probability	Integration
5.	Statistics	Definite integrals & applications
6.	–	The circle
7.	–	Set of complex numbers

The pre-service qualification for a secondary school teacher is a four-year B. Ed. degree offered by faculties of education at public universities. These faculties are responsible for developing both the content of the training and the accreditation of the process for teachers in secondary schools. As mentioned in section 1 above, the B. Ed. program in mathematics is presently offered in 10 faculties of education. It is significant to note that, at present, none of the academic staff of these 10 universities include a specialist in mathematics education.

The structure and content of the B. Ed. (honors) program in mathematics offered at the University of Khartoum is typical of corresponding programs at other universities. Students complete a total of 170 credit hours comprising 25 credit hours of university requirements, 50 credit hours faculty requirements and 95 credit hours of specialization (in this case mathematics) requirements. The mathematics requirements are comprised of 68 core credit hours: algebra (24), analysis (18), classical applied mathematics (12), probability and statistics (6), geometry (5) and numerical analysis (3). To complete the specialization requirements, a student must take one computer science course (3 credit hours) and choose 27 credit hours from a list of mathematics courses (30 credit hours) and physics courses (9 credit hours).

A total of 432 students enrolled in these programs in 2012–2013. Assuming a fixed rate of admission and graduation rate of about 80%, it is expected that higher education would produce about 346 qualified teachers of mathematics annually. To see this level of supply in proper perspective, note that the demand for mathematics teachers in Khartoum and River Nile state in 2011 was 835 and 325, respectively.

In-service training is coordinated by a general directorate, which is part of the Federal Ministry of General Education. It has been constrained for resources for a long time and, therefore, doesn't offer a regular program of in-service training. When some funds are available, it offers a program of basic training for new basic education teachers comprised of a total of 60 hours. Last year a similar program was offered for secondary school teachers in three states and it was comprised of a total of 90 hours.

Research in Mathematics and Mathematics Education

In view of the observations made above about few university faculty holding a doctorate degree in mathematics, I estimate that the total number of mathematicians working in Sudan is less than 50. An assistant professor is assigned a teaching load of 12 credit hours and, in view of poor levels of remuneration, almost all of them find themselves forced to engage in moonlighting in order to make ends meet. In addition, all university libraries in the country stopped subscribing to printed international journals in 1991. Thus, the environment in which these mathematicians work is not conducive for productive research in mathematics.

As for the number of doctorate degree holders in mathematics education in Sudan, there is only one now and the total number never exceeded three.

Table 6. Sudan: Number of internationally-visible publications in mathematics and mathematics education, 2000–2014

	2000	2002	2004	2006	2008	2010	2012	2014
Mathematics	1	1	2	4	3	4	12	24
Mathematics education	–	–	–	–	–	–	–	–

Source: SCImago. (2007). SJR — SCImago Journal & Country Rank. Retrieved October 29, 2015, from http://www.scimagojr.com

Concluding Remarks

Conscious efforts at building mathematics in most developing countries are relatively recent, beginning barely half a century ago. As is clear from the account presented in previous sections, Sudan's business of building a vibrant community of mathematicians remains largely unfinished. The list of priorities of policy makers does not include the development of mathematics nor mathematics education. In particular, it appears that the organic link between the promotion of mathematical research and mathematics education is not appreciated sufficiently. The onus of responsibility for bringing this about lies on Sudanese mathematicians. As stewards of their discipline, mathematicians need to concern themselves with school mathematics and its teachers. All institutions offering bachelor degree programs in mathematical sciences, need to assign high priority to teacher preparation, mathematics curriculum development, content-based professional development, partnerships with mathematics educators, and increased participation in the mathematics education community.

Mathematicians working in Sudan have recently established the Sudanese Mathematical Society. The business of promoting mathematics in the country should be the main item on the agenda of the Society. By initiating and coordinating appropriate activities, engaging mathematics teachers, learning from relevant experiences in the region and internationally and establishing appropriate partnerships, the Society can achieve a great deal.

Acknowledgements

The preparation of this chapter benefited from important information provided by Mohsin A. Hashim, University of Khartoum and Mohamed A. El Jack, Center for Curriculum and Educational Research and from discussions I had with both of them. Mostafa Abdin provided me with rare copies of Durell's trio. The inputs of all three are gratefully acknowledged.

References

African Development Bank (2012). African Economic Outlook: Sudan 2012. (Accessed on 23 February 2013 from http://www.youthpolicy.org/national/Sudan_2012_Youth_Unemployment_Briefing.pdf).

Black, P., & Wiliam, D. (1998). Inside the back box: Raising standards through classroom assessment. Phi Delta Kappan. 80(2), 139–148.

Ehrenberg, R. G., Brewer, D. J., Gamoran, A., & Williams, J. D. (2001). Class size and student achievement. Psychological Science in the Public Interest, 2(1), 1–30.

El Tom, M. E. A. (1979). Developing Mathematics in Third World Countries. North Holland.

El Tom, M. E. A. (2006). *Higher Education in Sudan: Towards a New Vision for a New Era.* Khartoum: Sudan Currency Printing Press.

Federal Ministry of General Education (2012). Educational Statistics 2011–2012. Khartoum.

MoFNE (2012). "Interim Poverty Reduction Strategy Paper." Ministry of Finance and National Economy, Khartoum. www.imf.org/external/pubs/ft/scr/2013/cr13318.pdf.

Nasell, I. (1985). *Hybrid Models of Tropical Infections*. Springer Verlag.
Schweigman, C. (1986). *Operations Research Problems in Agriculture in Developing Countries*. Khartoum University Press and Tanzania Publishing House.
SHHS (2010). Sudan Household Health Survey, Second Round, National Ministry of Health and Central Bureau of Statistics, Khartoum
World Bank (2012a). "The Status of the Education Sector in Sudan." Washington D.C. http://dx.doi.org/10.1596/978-0-8213-8857-0.
World Bank (2011a). "A Poverty Profile for the Northern States of Sudan." World Bank, Washington D.C. http://siteresources.worldbank.org/INTAFRICA/Resources/257994-1348760177420/a-poverty-profile-for-the-northern-states-of-sudan-may-2011.pdf.
World Food Program (WFP) (2015). WFP Sudan. Retrieved http://documents.wfp.org/stellent/groups/public/documents/ep/wfp269065.pdf.

About the Author

Professor Mohamed El Tom is Vice-Chancellor, Garden City University, Khartoum and Professor of Mathematics. He received his B.Sc. from Leeds University, UK and his D.Phil from Oxford University, UK. He was at various times mathematics faculty member in universities in Sudan, UK, Qatar and USA.

El Tom initiated the establishment of the School of Mathematical Sciences, University of Khartoum. Also, he initiated the establishment of the Pan-African Center for Mathematics as a collaborative project between Stockholm University, Sweden and the University of Dar es Salaam, Tanzania. El Tom chairs the Board of the Center. He is Fellow of the African Academy of Sciences and a member of the Reference Group for the International Science Program in Mathematics, Uppsala University, Sweden (since 2003).

In addition to his research interests in numerical analysis and mathematics education, El Tom published a book on higher education (in English and translated into Arabic) and edited two conference proceedings volumes on higher education.

El Tom chaired the African Mathematical Union Commission for Mathematics Education (1995–1999) and was a member of the Core Committee of the International Programme Committee of the 5th

International Conference on Mathematical Education (1981–1984). El Tom was member: International Program Committee for the International Commission on Mathematical Instruction Study on the Teaching and Learning of Mathematics at University Level (1997–2000).

El Tom loves all kinds of sports, particularly football (soccer). El Tom currently resides in Khartoum with his wife and their two daughters.

Appendix

Sample syllabus

SCHOOL CERTIFICATE
ALGEBRA

AN ALTERNATIVE VERSION OF
"A NEW ALGEBRA FOR SCHOOLS"

By

CLEMENT V. DURELL, M.A.

AUTHOR OF "GENERAL ARITHMETIC," "A NEW GEOMETRY FOR SCHOOLS,"
"MATRICULATION TRIGONOMETRY," ETC.

LONDON
G. BELL AND SONS LTD

PART II

https://doi.org/10.1142/9789813146785_0013

Chapter XIII

UAE: Brief Review of Education in the United Arabic Emirates

Edward G. Nichols
Zayed University

Anna Kohn
Teachers College, Columbia University

The UAE Education System

Overview

As part of wide scale educational reforms undertaken in the United Arab Emirates (UAE), all government schools have undergone a developmental overhaul over the past decade.

Because of its sudden abundance, the United Arab Emirates finds itself in the enviable position of being both an oil rich country and an economically developed state with sophisticated industries (CIA, n.d.). However, the windfall that came from the energy sector in the last few years is not something that is guaranteed to continue, and achieving great wealth in such an environment requires less human development than traditional industrialization, where wealth and sophisticated industries have to be built through ingenuity rather than simply drilled from the ground (Hamdan, 2012). To prepare itself for the time when oil is no longer the driving force in the economy, the UAE is reforming its education system to improve its population's skills. This is especially true for STEM (science, technology, engineering and math) subjects (Hamdan, 2012).

In analyzing the UAE education system, it is important to first discuss the political and social makeup of this new and enigmatic country. Firstly, the United Arab Emirates is a constitutional federation of seven emirates: Abu Dhabi, Dubai, Sharjah, Ajman, Umm al-Qaiwain, Ras al-Khaimah and Fujairah. The federation was formally established on December 2, 1971 (United Arab Emirates Chamber of Commerce, 2015). As the name implies, the UAE is a quasi-federalist system, with each emirate being roughly analogous to the states of the United States of America. Abu Dhabi is both the name of the capital city as well as the name of the largest emirate in terms of overall land area. It is also the possessor of the majority of the nation's oil reserves. Dubai is the largest city in terms of population and the second largest emirate (United Arab Emirates Chamber of Commerce, 2015).

This is a country in which the citizens, or Emiratis, make up only about 10–15% of the population. According to the World Bank's figures for 2013, the total population of the UAE was 9.3 million (dubaifaqs.com, 2015). However, these figures are always open to dispute, as accurate censuses are difficult to make, especially with the transient nature of the expatriate population.

Migration Policy.Org (Froilan T. Malit Jr., 2013), further breaks down the UAE's population into the following: As of "…2009, Emirati citizens accounted for 16.5% of the total population; South Asian (Bangladeshi, Pakistani, Sri Lankans and Indians) constituted the largest group, making up 58.4% of the total; other Asians (Filipinos, Iranians made up 16.7% while Western expatriates were 8.4% of the total population." The diverse and majority-expat population has led to the development of three different and distinctive education systems.

The most prestigious of the systems are the western style private schools. These are expensive and tend to serve the children of European, North American, and Diplomats from around the world. The more affluent Emiratis and expatriates from outside of Europe and North America also attend them. The government provides subsidies to help cover the cost for expatriates working in higher level jobs. The international private schools use an internationally accredited curriculum. This varies from school to school depending on its accrediting body, but most use the International

Baccalaureate Primary Years Program (PYP), the International Baccalaureate Program, (IB), the American Curriculum, or the British Curriculum (ADEC, n.d.).

Public schools exist but are only free for the local Emirati population. Others may attend for a fee. The government does not provide free education to any but its own citizens. However, the masses of expatriate laborers and lower-paid workers such as taxi drivers and restaurant employees are unable to bring their children to the UAE. Therefore, schooling is not an issue for them. They send money home to their countries of origin in order to pay for school for their children, if any, left behind there (Jandal, 2008).

For expatriates that are affluent enough to bring their families but not qualified to receive the school subsidy there is a third system. Either through subscription, or with the help of their home governments or NGO's from their home countries they set up their own private schools. For example, Pakistani businessmen and professionals have private, Urdu-speaking schools for their own children. They will use a curriculum similar to the one used in Pakistan. There are Indian private schools as well. They also use a curriculum adapted from their home country. In Abu Dhabi, ADEC offers the following 15 curricula at private schools: American, Bangladeshi, British, Canadian, French, German, Indian (2 types), International Baccalaureate (PYP, MYP and DP), Iranian, Japanese, Ministry of Education, Pakistani, Philippines, SABIS (ADEC, n.d.).

While The Federal Ministry of Education (MOE) and each local emirate's education authorities supervise all of these schools, the idea of school reform only applies to the Emirati Government schools, as the public schools are referred to (ADEC, n.d.).

More than most oil rich countries, the UAE has recognized the need for economic diversification. Its most cosmopolitan emirate, Dubai, has diversified its economy because it is projected to run out of oil faster than many other Gulf States (United States Energy Information Administration). Other emirates, including Abu Dhabi, have also begun diversifying their economies as part of an overall federal goal of lessening the country's dependence upon oil revenue. Part of the UAE's overall goal of diversification involves education reform, in particular focusing on teacher development and increasing the math and science skills of its population to better compete in the knowledge economy (Farah, 2012).

As part of the Gulf Cooperation Council (a trade block, sometimes military alliance, consisting of the UAE, Saudi Arabia, Bahrain, Oman, Qatar and Kuwait), the development of education has largely followed a script of massive investments in the ''hard'' infrastructure of education, e.g. physical schools and computers, rather than raising standards for how the students perform once they have left the system (McKinsey, 2007). McKinsey (2007) has called this the emphasis of inputs over outputs.

According to the UAE Embassy Washington DC (2009) education is vital to the UAE. Sheikh Zayed Bin Sultan Al Nahyan, the founder of the UAE, expressed, "The greatest use that can be made of wealth is to invest it in creating generations of educated and trained people." (Embassy of the United Arab Emirates Washington DC, 2009).

New programs are being initiated at all educational levels. A major area of emphasis has been to change Kindergarten through grade 12 (K–12) programs in order to enable UAE students to be empowered to attend international universities and compete in the global marketplace. Also, some of the finest higher education institutions in the world are opening branches in the UAE, attracting talented individuals in the Arab world and globally (Embassy of the United Arab Emirates Washington DC, 2009).

"Education reform focuses on improved readiness, greater accountability, better standards and improved professionalism (Embassy of the United Arab Emirates Washington DC, 2009)." Rote instruction is gradually being phased out with more interactive learning and English language training in math and science (Embassy of the United Arab Emirates Washington DC, 2009).

While each Emirate has its own local education authority, the Abu Dhabi Education Council (ADEC) in Abu Dhabi, the Knowledge and Human Development Authority (KHDA) in Dubai and the Federal Government's UAE Ministry of Education (MOE) are the leaders in the field of education reform. Abu Dhabi and Dubai have the greatest resources to spend and other emirates tend to either follow their leads, or rely on the Federal Ministry of Education for support in educational reform. All are tasked with carrying out reform, "while preserving local traditions, principles and the cultural identity of the UAE (Embassy of the United Arab Emirates Washington DC, 2009.)."

Secondary Education

According to stateuniversity.com (n.d.) secondary education in the UAE is based around the following sequence:

- Year I: Islamic Education, Arabic language, English language, history, geography, mathematics, physics, chemistry, biology, geology, computer science, physical education and family education
- Year II–III: Islamic Education, Arabic language, English language, mathematics, physical education and family education (for girls). Students are also allowed the ability specialize in a wide range of subjects in the liberal arts and sciences.
- Year III: philosophical subjects such as logic, psychology, sociology and economics are also taught.

Some of the objectives of education in the UAE are somewhat far removed from the objectives of most Western and East Asian forms of education. The stated objectives of the Ministry of Education are (United Nations Educational, Scientific and Cultural Organization, 2010):

- Inculcating faith in God and His prophets and moral values
- Inculcating Arab nationalism and pride in the homeland
- Learning the duties of citizenship and political and community participation
- Inculcating the value of work, production and perfection
- Contributing to comprehensive development and increasing technological achievement
- Developing methodical, critical and rational thinking
- Eradicated illiteracy
- Encouraging lifelong education

An additional priority of policymakers in the UAE is raising educational attainment for males. Dropout rates from secondary school for males in the UAE are around 25% (Hamdan, 2012). Only 30% of university attendees are males. There are a number of reasons for this.

Teachers in the UAE are not very engaging and male students are easily bored (Hamdan, 2012).

One of the factors that contribute to the high male dropout rates in the UAE has been the quality of teaching in mathematics classrooms. Because government secondary schools are all segregated by gender nearly all of the boys' teachers come from other Arab countries. They have been trained in the traditional teaching method, often referred to here as the Egyptian method, in which the teacher lectures and the student listens. In female classes, on the other hand, students are benefiting from the increased quality of teacher preparation programs because the vast majority of education students in UAE universities are female. Unfortunately, females are not allowed to teach males in government secondary schools.

One study conducted interviews with male school dropouts in the UAE, some of whom had left school as early as the sixth grade. In discussing why they had left school:

All interviewees agreed on two points about the school environment: bad teaching and an overwhelming amount of schoolwork. Feelings about bad teaching were based on teachers always reviewing material, class not being interesting and many times (for some courses) unable to be understood. "School was always the same thing. Teachers reviewing material in class and always pen and notebook must be available (Cruz, 2016).

Further exacerbating the problem, many men can drop out of school and still find government jobs that pay well or join family businesses. Because of how easy it is to find a job without an education, males drop out of school more readily (Hamdan, 2012). The government employs about 90% of UAE citizens (Hamdan, 2012). There is such wealth in the UAE that some students even come to school with assistants to help them carry bags (Hamdan, 2012). Though the trend is exaggerated in the UAE, the level of female achievement reflects a trend around the world (Hamdan, 2012).

Finally, much of the instruction in government schools is done in both English and Arabic. However, the instructional quality varies from school to school and from teacher to teacher. Therefore, because many students must learn math in a foreign language in which they are decidedly inexpert, they are unable to perform well on assignments and exams (Duval, 2000), (Duval, 2005).

STEM Education

The UAE Vision 2021 states that productivity and competitiveness are driven by "investment in science, technology, research and development" (United Arab Emirates Government, 2009, p.18) and can only be increased by prioritizing these areas and putting greater financial investments towards graduate programs and research, increasing enrollment rates in STEM (science, technology, engineering and mathematics), as well as building better relationships between academia and industry.

The UAE has a great need for STEM teachers in the private sector. Jobs for STEM teachers are readily available in international schools, vocational colleges and universities in Dubai and in other locations across the UAE. The UAE government is making strides to improve STEM education across the country to ensure the competitiveness of its economy and citizenry (PWC, 2015).

According to the International Study center Emirati fourth graders scored 434 on Trends in International Mathematics and Science Study tests, below a benchmark average of 500. Eighth graders scored 456 (Shabandri, 2012). Policymakers in the UAE recognize this problem and are putting more emphasis into teacher training (Farah, 2012). Governments of the various Emirates are trying to copy the models of those countries that scored the best on international math tests, such as Singapore, Japan, South Korea, Hong Kong and Finland (Farah, 2012). The Finnish model is of particular interest to the United Arab Emirates (Farah, 2012). This is because though the UAE has been able to train a vast quantity of teachers, it has had trouble improving the quality of teachers (Farah, 2012). Finnish standards for teacher training are very high and try to balance centralization and decentralization of education (Farah, 2012).

International Recognition

The student teacher ratio in Gulf Cooperation Council (GCC) countries is 12:1 (McKinsey, 2007). This compares with an average of 17:1 in Organization for Economic Cooperation and Development (OECD) countries and 24:1 in Singapore, one of the world's top scorers on international tests (McKinsey, 2007). Classroom size is often cited as a marker of a good school; however, these studies seem to indicate that teacher quality matters more than classroom size. The UAE has long pursued a policy of teacher quantity over teacher quality and it is now changing. Abu Dhabi has entered into a partnership with Singapore's National Institute of Education, one of the world's best teacher training programs (McKinsey, 2007).

Testing is another area where GCC countries, including the UAE fall short. Tests in the UAE mostly focus around memorizing information rather than how to apply it. Hierarchy is very rigid in the UAE, and a school principal who adheres to the rules set by the government will have higher ratings than one who's school is less rigid and more experimental in trying to improve outcomes (McKinsey, 2007).

McKinsey (2007) goes on to recommend that the functions of setting policy should be removed from the task of operating schools.

Perhaps one of the most exciting developments in education (and culture) in the United Arab Emirates is New York University's (NYU) Abu Dhabi campus. Very generous grants from the government were used to create the campus, which is something of a life apart from the rest of the country (Miller, 2013). Most students receive heavy financial assistance from generous scholarship program (Economist, 2015). The school resembles a luxury hotel to some extent, with fitness facilities full of personal trainers, video games and entertainment centers and free scuba lessons (Miller, 2013) (Economist, 2014).

Though NYU's Abu Dhabi campus' mission is primarily academic, one of the reasons for the luxury atmosphere has to do with the prohibitions on NYU students who go there. Many are internationals, including Americans, and the Emirate does not want them engaging in sensitive activities, such as traveling to nearby unstable hot spots in Lebanon, Israel or Jordan, and the university will actually pay them not to go (Miller, 2013). With the UAE trying to make itself more internationally and culturally prestigious through projects such as NYU, the Guggenheim and the Louvre, all located near each other, it would be very bad press if a student were to get hurt in the political upheaval in the Middle East. The partnership with NYU is in line with NYU President John Sexton's visions of the university as an internationally integrated institution (Miller, 2014).

Another special program includes the opening of The Sorbonne in Abu Dhabi in 2006, which will award qualifications under French regulations and standards set by the Sorbonne in Paris (Embassy of the United Arab Emirates Washington DC, 2009).

Final Remarks

The leaders of the United Arab Emirates are engaged in very long term thinking, especially when it comes to diversifying their country's reliance on oil. By drawing on the best talent and policies that successful nations and institutions abroad have to offer, they are on a very positive trajectory to effectively reform their education system through investments in math and science and make drastic improvements to their society and economy. Cooperation with institutions such as Singapore's NIE, Sorbonne and NYU, will give the UAE to compete in a future where oil is a less reliable source of economic growth.

References

Abu Dhabi Chamber (n.d.) Retrieved on April 30, 2015 from http://www.abudhabichamber.ae/English/AboutUs/Pages/About-UAE-Country.aspx.

ADEC. (n.d.). Adec.ac.ae. Retrieved April 26, 2016, from www.adec.ac.ae/en/Parents/PrivateSchools/Pages/Irtiqaa.aspx.

ADEC. (n.d.). adec.ac.ae. Retrieved April 26, 2016, from Abu Dhabi Education Council: https://www.adec.ac.ae/en/Education/OurEducationSystem/Pages/default.aspx.

ADEC. (n.d.). adec.ac.ae. Retrieved April 26, 2016, from Abu Dhabi Education Council: https://www.adec.ac.ae/en/CommunitySurvey/Pages/FAQ.aspx.

Barber, Mickael et.al. (2007). Improving education in the Gulf. The McKinsey Quarterly 2007 special edition. Retrieved on April 25, 2015 from https://abujoori.files.wordpress.com/2007/04/improve-gulf-education.pdf.

Central Intelligence Agency. (n.d.). Retrieved on April 15, 2015 from https://www.cia.gov/index.html.

Cruz, Y. (2016). Assisting Emirati Males to Remain in School Until 1 Graduation. Abu Dhabi: unpublished masters thesis.

Dubaifaqs.com. Retrieved on 2015, November 15 from http://www.dubaifaqs.com/population-of-uae.php.

Duval, R., Ferrari, P. L., Høines, M. J., & Morgan, C. (2005). Language and Mathematics. Language and Mathematics, 789.

Duval, R. (2000). Basic issues for research in mathematics education, in T. Nakahara & M. Koyama (Eds.), Proceedings of the 24th Conference of the International Group for the Psychology of Mathematics Education (Vol. 1, pp. 55–69), Hiroshima University: Hiroshima, Japan.

Embassy of the United Arab Emirates, Washington DC, (2009). Retrieved on April 28, 2015 from http://www.uae-embassy.org.

Farah, Samar. March 2012. Education Quality & Competitiveness in the UAE. Al Qasimi Foundation. Retrieved on April 30, 2015 from http://www.alqasimifoundation.com/Files/Pub3-paper(Samar.Farah).pdf.

Froilan T. Malit Jr., A. A. (2013, September 18). Labor Migration in the United Arab Emirates: Challenges and Responses. Retrieved from Migration Information Source: http://www.migrationpolicy.org/article/labor-migration-united-arab-emirates-challenges-and-responses.

Hamdan, Sara. (2012, May 27). In the Gulf, Boys Falling Behind in School. The New York Times. Retrieved on April 28, 2015 from http://www.nytimes.com/2012/05/28/world/middleeast/28iht-educlede28.html.

About the Authors

Edward Nichols has been an Assistant Professor at Zayed University in The United Arab Emirates since 2012, where he has been involved in designing and implementing teacher training programs and educational reform programs in Abu Dhabi. Before that he taught in teacher training programs at the University of Arizona in Tucson for 4 years. He received his PhD in Reading, Language and Culture from the University of Arizona in 2009. Prior to moving into the realm of academics he spent 25 years teaching first and second grade in a borderlands elementary school in Arizona in the USA. He has considerable expertise in early childhood education, including math and literacy, and developed innovative math education programs for use in multiage classrooms. His chief research interests include multimodality in mathematical texts, the pedagogy of play, and effective teacher training in an English as a Foreign Language Environment.

Anna Kohn is currently a PhD student studying Mathematics at Teachers College, Columbia University. She obtained her bachelor's and master's degrees in Applied Mathematics and Physics from Moscow Institute of Physics and Technology. Anna Kohn has extensive experience teaching mathematics. She has worked as an Adjunct Professor of Mathematics at University of Bridgeport and Westchester Community College, as well as a math teacher at a public school and a Teacher's Assistant for Computational Mathematics at Moscow Institute of Physics and Technology. Anna's academic interests include but are not limited to: the mathematical giftedness of school children, online mathematics education, and problem solving. She has published an article "Informal methods of identification of mathematical giftedness of elementary school students."

https://doi.org/10.1142/9789813146785_bmatter

Epilogue

There are currently 48 Muslim-majority countries that are often considered to form the Muslim World. Diversity is a salient feature of this world. In view of this and the fact that the education systems of Muslim countries followed different trajectories as amply detailed in this anthology, it is natural to observe that diversity also characterizes the respective landscapes of mathematics and its teaching in these countries. However, these countries' mathematical developments share in common several significant features.

The continued implementation of a teacher-centered approach to teaching mathematics, emphasis on rote learning and adoption of summative assessment as the only form of evaluating students' learning, are examples of such features. Also, few universities, not only in the countries covered by this anthology but also in the Muslim world as a whole, offer postgraduate programs in mathematics education. Another feature which is shared by the Muslim-majority countries is the low level of research output in mathematics education. Certainly, these features impede the efforts to improve mathematics education and need to be addressed.

Examples of common features that help promote mathematics education include efforts at institutional building (such as mathematics and teachers of mathematics associations, mathematics journals and clubs), the almost uninterrupted interaction between national experts and their international counterparts in mathematics education through regional and international conferences/workshops which provide information on trends and best practice in mathematics education. Also, the active involvement of university departments of mathematics in curriculum reform, the organization of national and regional meetings that discuss issues of teaching and learning of mathematics, and the writing of textbooks represent a significant departure from past experience in which academic mathematicians showed little concern for school mathematics.

Perhaps, the results of these collective efforts are most evident in the case of the Islamic Republic of Iran where they contributed to the remarkable and unprecedented achievement of Maryam Mirzakhani, the first woman to ever earn a Fields Medal. The talent of Maryam Mirzakhani was identified and carefully nurtured in her home country, Iran. It is not unreasonable to expect that her remarkable achievement to attract and inspire many young talents in mathematics, not only in Iran but also in the Muslim world at large.

Mohamed El Amin A. El Tom
The Ministry of Education, Khartoum

Printed in the USA
CPSIA information can be obtained
at www.ICGtesting.com
LVHW021745271223
767385LV00007B/642